A BIOMASS FUTURE FOR THE NORTH AMERICAN GREAT PLAINS

ADVANCES IN GLOBAL CHANGE RESEARCH

VOLUME 27

The titles published in this series are listed at the end of this volume.

A BIOMASS FUTURE FOR THE NORTH AMERICAN GREAT PLAINS

Toward Sustainable Land Use and Mitigation of Greenhouse Warming

by

Norman J. Rosenberg

Springer

A C.I.P. Catalogue record for this book is available from the Library of Congress.

ISBN-10 1-4020-5600-1 (HB)
ISBN-13 978-1-4020-5600-0 (HB)
ISBN-10 1-4020-5601-X (e-book)
ISBN-13 978-1-4020-5601-7 (e-book)

Published by Springer,
P.O. Box 17,3300 AA Dordrecht, The Netherlands.

www.springer.com

Cover photo; "Fleeing a dust storm". Cimarron County, Oklahoma Arthur Rothstein, photographer, April, 1936 (Library of Congress).

Printed on acid-free paper

This book is dedicated to
Joshua, Rachel, Ariella,
Daniel, Alyssa, and Bettina
and, of course to Sarah

TABLE OF CONTENTS

PREFACE

WHY THIS BOOK?

This book is an exploration of the possibility that a significant portion of the North American Great Plains (NAGP), now primarily in rangeland, corn, soybean, and small-grain production, can be converted to the production of biomass-energy crops. Biomass can be used as a substitute for some of the fossil fuels the use of which is now increasing the atmospheric concentration of carbon dioxide (CO_2) and contributing to global warming and climatic change. Such a land use change to biomass could lead not only to a global good but also to specific economic and environmental benefits for the NAGP region. This analysis is prompted by the following facts and trends:

> The emission of CO_2 from fossil fuel combustion and tropical deforestation and the rising concentrations of other greenhouse gases make global warming a virtual certainty in this century; indeed the evidence is strong that a warming is already discernible.
>
> Global warming will lead to climatic change and, while the geographic distribution of this change is not yet known, most general circulation models (GCMs) suggest that midcontinental regions in the northern hemisphere (such as the NAGP) are likely to become drier as well as warmer.
>
> NAGP, one of the world's major breadbaskets, is subject to periodic droughts and other climatic stresses that may worsen with global warming. Thus, it is prudent, if only for their own benefit, that the people and governments in this region seek ways of reducing the emissions of greenhouse gases.
>
> Among climate change mitigation strategies now under consideration are the expansion of nuclear power production, capture of fossil fuel carbon at the smokestack and its transport to and sequestration in geologic strata and the oceans, afforestation, introduction of substantial solar and wind energy infrastructure, sequestration of carbon in soils in the form of organic matter and production of biomass as a substitute for fossil fuels. Each of these options has associated physical, environmental, and economic risks. Soil

carbon sequestration and biomass production are among the most environmentally benign options and are well suited to the NAGP region.

Whether combusted for boiler-fuel or converted to liquid transportation fuels such as ethanol, biomass essentially recycles carbon, withdrawing CO_2 from the atmosphere through photosynthesis and returning it when the vegetation or its derivative products are consumed.

Whereas fossil fuels burden the atmosphere with carbon drawn from ancient geologic storages, biomass recycles carbon and should add little or no additional carbon to the atmosphere. Indeed, if a way can be found to capture the carbon released in biomass combustion or processing and to sequester it (as is actually being done on an experimental scale for coal-fired power plants), biomass could lead to a net "negative emission" of carbon and, perhaps, actually lower the atmospheric concentration of CO_2.

Ethanol is now produced in the NAGP and adjacent regions primarily from the starch in grain corn. Some calculations have shown that the ethanol so produced is nearly energy-neutral, its final energy content being not much greater (perhaps even less) than the energy required to produce it. Recent analyses suggest an energy gain of as much as 25%. For decades now, ethanol production for fuel has been profitable to producers only if directly or indirectly subsidized.

In addition to the use of high-cost grains, it is now feasible to convert the cellulose in corn stover, wheat straw, and all types of biomass to ethanol through the use of enzymes that break the strong bonds between sugar molecules that make up the cellulose structure. New enzymes, genetically engineered for that purpose, will make cellulosic ethanol far more energy efficient and cost competitive than ethanol from grain can ever be. However, $60–70 per barrel (and who knows how much higher the price may go?) petroleum makes even ethanol from grain appear competitive. Production from cellulosic biomass materials should be even more so. Unsubsidized ethanol production from biomass is conceivable.

It has been strongly argued that agriculture as currently practiced in the NAGP is unsustainable. Although soil erosion has been reduced by conservation strategies implemented since the "dirty-thirties," overuse of chemical fertilizers and pesticides and the mining of groundwater resources create serious environmental problems in this as in many other agriculturally intensive regions of the world. For these reasons some scholars have suggested that large portions of the NAGP should be returned to its native vegetation and fauna and that a "Buffalo Commons" replace its farms and ranches.

While the world faces a problem of consequential climatic change, it also faces population pressures and expected improvements in worldwide living standards that will increase demand for both energy and agricultural products. Thus, diversion of land to grass or biomass energy production raises the question of how needed food, feed, and fiber will be provided in the future.

Perennial biomass crops such as switchgrass (*Panicum virgatum*) and the poplar (*Populus* spp.) (as well as many other species being studied at this time) require less tillage and fewer pesticide and fertilizer applications and may, thus, be better sustainable than the traditional crops they could replace (as well as being more popular and politically feasible and likely of implementation than the Buffalo Commons). As the roots of crops—particularly perennials—decay, they deposit carbon in the soil, some of which is sequestered for very long periods of time. In addition, conservation tillage practices consistent with sustainable agriculture favor soil carbon sequestration. Such tillage practices require less energy than conventional farming practices and may also favor biodiversity. Biomass crops, which are mostly perennials, should prove better with regard to soil carbon sequestration than traditional annual crops. The culture of biomass crops is consistent with conservation tillage.

Each of the issues and questions posed above are addressed in this book as is the matter of whether ecological, environmental, and economic arguments support the need for a significant conversion of NAGP lands back to grass cover, to biomass cropping, or both. Two other questions addressed: Can genetic modification of biomass crops increase their productivity to the point of economic competitiveness? Can the productivity of traditional (or new) crops be increased sufficiently to compensate for decreases due to conversion of substantial areas of agricultural land to biomass production?

WHY THIS WRITER?

This book is a labor of love. Immediately upon completion of graduate studies in soil physics and meteorology at Rutgers University, I was hired by the University of Nebraska to initiate a program of research in agricultural climatology. The objective of this program was to find ways to diversify Nebraska's agriculture, then (as now) almost entirely devoted to extensive corn, soybean, and wheat production, to enable production of higher per acre value vegetables and industrial crops.

Nebraska has excellent soils and plentiful water resources, but its climate is severe. My job was to find ways to somehow reduce vulnerability of the high-value crops to climatic stresses or to protect these crops from them. My first experiments were aimed to improve understanding of how windbreaks could best be designed to moderate the microclimate of crops grown in their lee. After working for a few years on that problem (with some success) I began to study how the water needs of crops might be minimized by wind shelter, by timing of irrigation, and by modification of the plant's "architecture" (e.g., leaf distribution, leaf color, and light penetration into the plant canopy). In order to do the necessary field experiments I had to adapt existing instruments or develop *de novo* nondestructive methods to measure evapotranspiration and photosynthesis in the field. Hence I gained experience with micrometeorological sensors for measuring the instantaneous exchanges between soil, plant, and atmosphere of heat, momentum, water vapor, and CO_2. Many of these studies were carried out from a laboratory trailer that traveled around the State. But in the mid-1960s I was able to establish a "home-base" at the University of Nebraska Field Laboratory near Mead, Nebraska, on land that had been a part of a recently decommissioned armaments production facility. There my colleagues and I erected a permanent instrument tower for continuous year-round measurements of wind speed and direction, temperature, humidity, and CO_2 concentration.

In 1958, as part of the International Geophysical Year, sensors were placed at the top of Mauna Loa, a 4167 m (13,678 ft.) high volcanic mountain on the "Big Island" of Hawaii to document the changing global concentration of atmospheric CO_2 which was then rising at a rate of ~1 ppmv per annum. Mauna Loa had been chosen for this purpose because its altitude and remoteness from strong local sources and sinks for CO_2 allowed for monitoring a well-mixed atmosphere.

Our observations at Mead, at an altitude of 366 m (1200 ft) in the middle of a vast agricultural region, differed in important ways from those at Mauna Loa. The amplitude of the daily CO_2 concentration wave was far greater at Mead because of the daytime drawdown due to photosynthetic capture of CO_2 by the region's crops and its nocturnal rise due to respiratory release. Similarly, the annual wave had much greater amplitude because of net growing-season capture of CO_2 and its net release during the winter when the vegetation was either dead or dormant. But, the inexorable annual rise in the mean annual concentration was essentially identical at Mead on the eastern Great Plains and Mauna Loa in the central Pacific.

This observation piqued my interest in the entire question of climatic change—an interest that has dominated my career ever since. I began to speculate and write about how climatic change might affect agricultural productivity around the world, but particularly in the USA. I became convinced that there was at least one benefit for agriculture in the rising CO_2 concentration, i.e., that plants would grow bigger and faster because of increased rates of photosynthesis and would use less water in the process because CO_2 induces closure of the stomates (pores) on the leaf through which water vapor exits the plant into the air above. And, although I was convinced from the first principals of thermodynamics that climate must change with continued increase in atmospheric CO_2 concentration, I was skeptical then (and still am to a degree) about how well the regional distribution of changing climatic factors—temperature, precipitation, etc.—could be predicted.

In the late 1970s, I experienced yet another "Nebraska-epiphany" that altered the course of my career. The Great Plains region was struck by another of its very severe droughts in 1977. With others, I was called to the office of the then Governor of Nebraska, J.J. Exon, to advise on what state government should be doing to help the states' farmers and ranchers. I was struck at that meeting by the fact that, despite almost a century of university and agency research on various aspects and facets of drought, there existed no organized plan for dealing with that most inevitable of Great Plains phenomena. It became clear to me that a plan was needed to prepare for and cope with drought and that such a plan must consider not only climatology and agricultural research, but also the societal, economic, political, and even psychological, impacts of drought as well. Having raised the question I was of course charged with setting the process in motion. The outcome of the planning process was *A Drought Strategy for Nebraska*. This concept of drought strategy has been carried forward by others and much more sophisticated and effective plans have by now been developed for most states and many nations as well. But my interest in drought—after death and taxes the third inevitability (at least on the Great Plains) remains strong and has taken me to many other drought-affected regions of the world.

My interest in climate change and drought led in 1987 to a job change. I joined Resources for the Future, a Washington, DC-based "think thank" to develop a climate resources program. A major product of that program was a study of the potential impacts of climatic change on the agriculture, water resources, forests, energy, and overall economy of the central US region comprised of Missouri, Iowa, Nebraska, and Kansas (the MINK study). The research had been funded by the Department of Energy and done in cooperation with the Pacific Northwest National Laboratory. In 1992, I was invited to join Pacific Northwest National Laboratory (PNNL) where I continued to study impacts of climate change on US and world agriculture and water resources.

From the 1970s to this time the evidence has strengthened that mankind is altering the earth's climate through emissions of CO_2 from fossil fuels and other greenhouse gases and by tropical deforestation, burning, and other forms of land use change. By this point, I am fully convinced that anthropogenic climatic change is real and that steps must be taken to reduce the rate of change and, hopefully, reverse its direction.

There is within PNNL a strong program to develop technologies that can contribute to the mitigation of climatic change by finding ways to reduce greenhouse gas emissions and/or to capture and sequester the CO_2 that continues to be emitted. In the final five or so years of my pre-emeritus career at PNNL I participated in that program—termed the Global Technology Strategy Project (GTSP)—and helped in developing an understanding of how two mitigation technologies, both focused on agriculture, might contribute to the overall goal. These technologies are soil carbon sequestration and biomass to substitute, at least partially, for fossil fuels.

This book, then, is a synthesis of my ideas on how climate change might affect the US and Canadian Great Plains, a region that continues to be of great interest to me, and how the Great Plains can contribute to a solution of the climate change problem through the two agriculturally based technologies mentioned above. Again, then, this book is a labor of love—for a region, for the physical and social sciencesas well as scientists with whom I have been privileged to work, for our planet, and for future generations threatened by anthropogenic climate changes whose long-term consequences we may know only after much damage has been done.

ACKNOWLEDGMENTS

In writing this book I have had the assistance and encouragement of many colleagues. Although this list will be, I hope, exhaustive, inevitably some deserving names will be neglected. I hope to be forgiven for this.

First I wish to acknowledge my colleagues in the Joint Global Change Research Institute (JGCRI) in College Park, Maryland, and other units of the Pacific Northwest National Laboratory, Richland, Washington. These include my closest associates, Cesar Izaurralde and Allison Thomson, who provided advice and graphics on soil carbon and climate change effects on US dryland and irrigated agriculture and water resources. At my request James Dooley and Casie Davidson prepared a map of the geological strata underlying the Great Plains and adjacent areas where captured carbon dioxide might be sequestered. Jae Edmonds, Hugh Pitcher, and Steven Smith of JGCRI directed me to specialized databases needed for this work. Paulette Land provided good advice and help on computer manuscript preparation.

Next, I thank the colleagues and friends from my days at the University of Nebraska, Lincoln for the guidance they have provided. These include Professors Shashi Verma on soil carbon sequestration experiments, Charles Francis on sustainable agriculture in the Great Plains, and James Brandle on windbreaks as a source of biomass in the region. Special thanks are also due to Professor Atul Jain of the University of Illinois Urbana–Champaign for a "commissioned" run of his soil carbon sequestration model. Professor Bruce McCarl of Texas A&M University and William Cline of the Institute for International Economics shared manuscripts in preparation and published papers on the economics of biomass for fuel and soil carbon sequestration.

Thanks are extended to Drs. Hari Eswaran and Craig Ditzler of the US Department of Agriculture (USDA) Natural Resources Conservation Service who prepared the map covering soils of the Great Plains region. Dr. Raymond Motha (a former student, now Chief Meteorologist of the USDA World Agricultural Outlook Board) provided data on crop production trends over time. Thanks, too, to Drs. Kenneth Vogel—USDA Agricultural Research Service, Lincoln, Nebraska—Stan Wullschleger, Sandy McLaughlin, and Robin Graham of the Oak Ridge National Laboratory, and Zia Haq of the Department of

Energy/Energy Efficiency and Renewable Energy Division, all of whom provided information and data regarding the genetics of biomass crops, their processing, and related environmental benefits and concerns. Another group of biomass researchers including Dr. John Berdahl of the USDA Agricultural Research Service, Mandan, North Dakota, Dr. Arnold Kruse of North Dakota's Fish and Wildlife Agency (retired), and Professors Arvid Boe, Kevin Kephardt, and Vance Owens of South Dakota State University provided much needed information and insight about the performance of candidate biomass species for the northern and western portion of the Plains.

Thanks are due to institutions as well as people. With support of the US Department of Energy/Office of Science, under its Global Technology Strategy Program led by Dr. Jae Edmonds, my home institution, the JGCRI, provided the services of a research assistant for most of the 2005–2006 academic year. Without David Daddio, University of Maryland economics student, diligent scholar and fact-checker, this project would have taken considerably longer than it has (and that's long enough!).

I also thank the Rockefeller Foundation which supported a month-long residency for me and my wife, Sarah, at its Study Center in Bellagio, Italy, where the final touches were put on the manuscript of this book.

Finally, I thank Sarah for bringing to this effort keen interest in the subject and the instincts of an experienced researcher and historian, and, not least, for her loving support, patience, and forbearance.

CHAPTER 1

INTRODUCTION

1. IMAGES OF THE PLAINS

A review in the *Washington Post* of Michael Cook's *A Brief History of the Human Race*[1] drew my attention to this quotation:

> The basis of farming and hence of the whole historical development of human societies, is grass. ... This immediately explains why the tropics have taken a back seat in the course of history (even today underdevelopment has an elective affinity for the tropics). The action is where grass is; and grass despite its tropical origins, is most successful in temperate climates.

Whether correct or not (and some of its contentions seem questionable) Cook's phrase "the action is where grass is" seems a good place to begin this book, an examination of future prospects for one of the world's most extensive and extensively developed grasslands—the Great Plains of North America.

Yet another quotation describes the ambience of the Plains in a telling way. The critic Peter Schjeldahl reviews the work of Agnes Martin,[2] a nonagenarian artist whom he describes as "from Saskatchewan, up north on the tabletop of the Great Plains." Her "spare ascetic abstractions," Schjeldahl explains, are influenced by the Plains:

> There is nothing cuddly about nature in that neck of the non-woods, where vicious cold and exhausting heat, ceaseless wind, and, alternating underfoot, snow, ice, sucking mud, and black dust try the soul. By way of compensation there's sky. A tremendous inverted bowl bells down over every horizon—affording far-distant glimpses of other people's weather. The god of the plains is an orthodox minimalist, specializing in brute coups of uninflected space and light. Checkerboard roads and evenly distributed granary towns—mapped

[1] Norton (2003). Review by Michael Dirda, "Sunday Book Reviews," *Washington Post*, November 2, 2003, p. 15.

[2] Peter Schjeldahl. "Life Work: Two Shows from Agnes Martin," *New Yorker*, June 4, 2004. p. 94.

> in advance, at the time of settlement, by governmental and railroad bureaucrats—advertise humanity as a complementing force of sublime heartless logic. A sense of existence as seamless and intractable—all one hard thing—crushes and exalts the plains dweller, inducing both humility and lofty thoughts.

However that may be, the grass cover and the landscape of the Great Plains have surely influenced its peoples and their way of life.

2. GRASSLANDS

Grasses give the Great Plains its character. Prior to European settlement the bulk of the biomass in the native vegetation of the Plains was composed of grasses in which other herbs and shrubs grew. Most grasses in the Great Plains are perennials. Vinton (2004) describes their species distribution as follows: Tall grasses grew in the eastern more moist portions of the Plains. The dominant species were big bluestem (*Andropogan gerardii*), switchgrass (*Panicum virgatum*) and Indian grass (*Sorghastrum notans*). Short grass prairie typified the western regions where rainfall is more limited. The dominant grasses were blue grama (*Bouteloua gracilus*) and buffalograss (*Buchloe dactyloides*). Mixed grass prairie lay between the tall and shortgrass prairies. Grasses typical of both bordering regions appear in the mixed prairie as well as such species as little bluestem (*Schizachyrium scoparius*) and western wheatgrass (*Agropyron smithii*). The northern portions of the mixed grass and tallgrass prairies extend into Canada and include more of the cool-season species. There is also an area in the Canadian Prairie Provinces dominated by rough fescue (*Festuca scabrella*). The southern portion of the mixed grass prairie is dominated by the warm-season grasses—little bluestem and side-oats grama (*Boutelova curtipendula*).

As Vinton explains, the dense fibrous root system of these grasses and the annual dieback of the aboveground portions of the grasses constantly enrich the soil with organic matter, adding nutrients and increasing its water holding capacity. Additionally, grasses persist under drought, grazing and fire. Because of regenerating underground organs that live over from season to season and because of large quantities of nutrients and energy-containing compounds stored in the root system, grasses can recover when their aboveground portions are killed or removed by grazing. The soils that developed under this vegetation are the foundation of Great Plains agriculture. Grazing on the untilled land has changed both the mix of species on the range and its productivity.

3. “DISCOVERING” THE PLAINS

3.1. Pre-European

If it can be said that the Great Plains was “discovered” it was, of course, by the ancestors of the people we in the USA now call “Native Americans” or, in Canada, members of the “Original Nations.” Their immensely rich history and accomplishments cannot be done justice in this work since, limited by necessity, this book deals, essentially with land use in the Plains after exploration and settlement by people of European origin. Prior to the early appearance of Europeans on the Plains, Native American tribes had lived mostly along streams in semipermanent settlements. Because they lacked the means of rapid long-distance movement, the Indians were not always able to reach the migrating herds of buffalo which were their primary source of food. That is, until horses, reintroduced to the region by Spanish explorers of the Southern Plains, became available to them in the early 17th century. By the early 19th century, when American settlement of the region was beginning, Indians roamed the region extensively, able to freely follow the buffalo migrations.

3.2. Early explorations

Early European-origin explorers of the region adopted very different views of the Plains. Reports of the Lewis and Clark Expedition (1803–1806) described many areas lacking all timber and a scarcity of water. The far northwestern corner was described as a "desert." Zebulon Pike crossed from the Missouri River to the Rocky Mountains through what is now Kansas in 1805–1806 and in a report to the War Department described that region as a "desert." "These vast plains of the western hemisphere may become in time as celebrated as the sandy deserts of Africa," Pike's report stated.

Stephen Long crossed what is now Nebraska with his expedition of 1819–1820. His reports provided little to contradict Pike's earlier findings. Indeed, a map drawn by the Long Expedition's cartographer labeled the Missouri and Arkansas basins as "Great Desert"—a notion that persisted through much of the 19th century. Dick (1975) reports on accounts by succeeding travelers—virtually all of which (despite much evidence to the contrary) reinforced the notion of the Great Plains as a desert.

The Canadian portion of the Plains also received "unfavorable reviews" from its early and influential surveyors, John Palliser and Henry Hind (1857–1860). Palliser described the region as "...sterile with scanty pasturage"—an extension in a sense of the "Great American Desert."

3.3. Settlement

Despite these reports, by the 1850s settlements were beginning to appear in Nebraska on the western side of the Missouri River and at the eastern edges in what are now the eastern Great Plains states. Little by little the "border" of the desert pushed westward.

Encouraged by the Homestead Act of 1862 which offered free land to early settlers and by the railroads which had received extensive land grants along their rights-of-way, the farming frontier moved westward more rapidly. Reports of excellent crops on the newly broken prairies further weakened belief in the existence of a Great American Desert. The Homestead Act provided settlers with 160 acres of land and a mule. Settlers were required to live on the land for 5 years and improve it before they could gain title. At least some portion of their homestead was to be cultivated. In the humid eastern regions of the country the acreage provided was more or less sufficient. In the semiarid west, however, 160 acres (~65 ha) proved too little to support a family.

The 1870s were a period of higher than normal rainfall. Boosters urged more settlers to move into the region. Even the scientific establishment of the time speculated that settlement had changed the climate for the better. One theory cited by Dick (1975) proposed that settlement would reduce the numbers of prairie fires and as a result more trees would grow which would in turn increase rainfall. Another theory proposed by Professors Charles D. Wilber and Samuel Aughey of the University of Nebraska was summed up in the expression "rainfall follows the plow." As Aughey explained: "It is the great increase in the absorptive power of the soil, wrought by cultivation, that has caused, and continues to cause an increasing rainfall in the State." According to this theory, the surface of unplowed prairie soil is compact from the treading of countless buffalo and other animals and absorbs little of the falling rain; most runs off to the nearest streams. By contrast, plowed soil, like a sponge, is open to absorbing all of the rain that falls and, through evaporation, slowly releases that moisture back into the atmosphere where it makes rain. This theory, put forth by my distinguished predecessors at the University of Nebraska, was quite misinformed. Careful observations show that where soil is vegetated its permeability remains high. The aggregated crumbs of newly tilled soil, if uncovered for long, are quickly broken down by heavy rains and the surface is compacted and/or crusted-over. Bare soil in such condition is eroded as water runs off the land.

The notion that "rainfall follows the plow" gave boosters a strong argument to encourage settlement and further westward advance of the farming frontier. The generally good rains of the 1870s and 1880s encouraged a great rush across the 100th meridian with some counties populated in a year or two. Ranchers, however, continued to argue that only grazing was sustainable (not then a "buzzword") in the western reaches of the Plains.

A more valid "scientific" assessment of prospects for the Plains region was that of John Wesley Powell in his *Report on the Lands of the Arid Region of the United States* published in 1879. In a sense his findings favored the western rancher's view by recommending that west of the 20 inch (~500 mm) rainfall belt—roughly the 100th meridian—land should be used only for grazing, irrigated farming, or semiarid cropping.

In the late 1880s and almost continuously in the 1890s nature, unconvinced that "rainfall follows the plow," returned to more typical patterns and then to a severe and protracted drought. Settlers were forced to move back to eastern portions of the Great Plains states and even further east. Historian Henry Nash Smith (1950) described the settlement patterns on the Plains between 1870 and 1890 in this way: "settlement advanced far out upon the Plains in periods of relatively high rainfall, only to be forced back by the dry periods which always followed." Dick (1975) also writes of "...a surge of settlement cut short in 1874 by grasshoppers and drought."

The Canadian portion of the Plains was settled in a somewhat more deliberate manner than the American portion. Prior to 1869, when it was sold to the new Dominion of Canada, the region belonged to the Hudson's Bay Company. The government's objectives were to people the Prairie Provinces with farm families, to connect the region to central Canada with a railroad and to exploit the region's resources (Thompson, 2004). The Dominion Lands Act of 1872 was modeled after the US Homestead Act, requiring that farmers plant their 160 acres, build a home (even if only a shack) and survive on the homestead for 3 years. As in the USA, private railroads were subsidized by government and granted large areas of land. The path of the railways influenced settlement patterns in Canada as it did in the States. European settlement of the Canadian Plains began in earnest, however, only in the first decades of the 20th century.

The US Congress attempted to remedy deficiencies in the 1862 Homestead Act with regard to the western Plains. The Timber Culture Act of 1873 aimed, because of their supposed benefits, to promote tree-planting in the treeless areas of the West. It allowed settlers to expand their holdings if they planted and maintained trees for a certain period.[3] The program proved ineffective and the law was repealed in 1891.

Next came the Kinkaid Act of 1904. This was at first a special homestead law which applied only to the western and central portions of Nebraska (primarily the Sand Hills). The act allowed for larger homesteads in the designated areas, except for lands set aside as being suitable for irrigation. The act was an effort to respond to the fact that 160-acre tracts were far too small for productive agriculture and ranching in the relatively arid Sand Hills and high plains regions of Nebraska. The Kinkaid Act allowed acquisition of one section (1 sq. mile: 2.6 sq. km), equal to 640 acres (~260 ha).

The settlement policies of Canada and the USA differed in one important way. Canada allowed certain groups like the Mennonites to live in "hamlets," away from their land and yet earn title after farming it for 3 years. Other groups such as Icelanders, Mormons, Jews, Ukrainians, and Hutterites were drawn to the Prairie Provinces. Prosperity based on good crops prompted continued immigration from Eastern Canada, Europe, and the USA in the early decades of the 20th century.

4. THE DUST BOWL YEARS

The cycle of settlement in and after good years followed by out-migration during and after bad years continued, with most severe consequences when drought inevitably returned to the Great Plains in the 1930s.

The terms "Dust Bowl" and "Dust Bowl era" are properly applied to the southern Plains region covering southwest Kansas, eastern Colorado, northern

---

[3] Nebraska State Historical Society, US Government Land Laws in Nebraska, 1854–1904. (http://www.nebraskahistory.org/lib-arch/services/refrence/la_pubs/landlaw7.htm)

New Mexico, and the Texas and Oklahoma Panhandles. The term "Dust Bowl" is, however, often associated in the popular mind with the entirety of the Plains region. Overcultivation and overgrazing, especially in the southern Plains had exposed soil surfaces, making them susceptible to wind erosion under conditions of drought. The rapid expansion of wheat production during and after World War I contributed to the increased vulnerability of the region to dust storms. Hurt (2004) describes how the Dust Bowl began with drought in fall of 1931 ruining the wheat crop so that by late January of 1931 the prevailing winds began to lift the soil setting off dust storms.

The dust storms and the concurrent economic depression that began in October 1929 created conditions that required federal government response. Programs were developed to provide funds to farmers agreeing to limit wheat and cotton production as well as price supporting loans for these crops. The Soil Conservation Service encouraged adoption of erosion reducing techniques such as contour terracing, grass water ways, strip cropping and others, and provided financial assistance to enable farmers to make the necessary changes. Wind-eroded land was purchased by the government and returned to grass cover; these areas are now known as National Grasslands. These and a host of other governmental interventions played a very important role in stabilizing Plains agriculture and providing a "safety net" for its farmers. As Wilhite et al. (1986) have shown, these and follow-on programs have considerably moderated both the physical and societal impacts of subsequent droughts on the Plains.

It is also apparent that government resettlement policies were influential in encouraging out-migration from the southern Plains Dust Bowl. But according to Bonnifield (1979) many more people "toughed-it-out" than actually abandoned the region. This he attributes to the fact that the original settlement process was a difficult one and hardened the populace. And the experience of good years and bad had also imbued in them a realistic expectation of what life in the region had to offer. It was the coupling of the two drivers—drought and economic depression—that explains the extensive out-migration from the Great Plains during the 1930s.

Lessons of great consequence for the region were learned during the Dust Bowl years. The severity of the drought in the early 1930s prompted a strong and, on balance, effective response from the US Federal Government. Some programs were already in effect by 1936 when the *Report of the Great Plains Drought Area Committee* was submitted to President Franklin D. Roosevelt (Cooke et al. 1936).[4] The aim of this high-level committee was to "outline a long term program which would render future droughts less disastrous" and to assure the "most efficient utilization of the natural resources of the Great Plains area." By the time of report issuance the federal government had already expended some $475 million on drought related conservation works as well as grants, loans, and

[4] These programs included the Resettlement Administration, Civilian Conservation Corps, Works Progress Administration, Agriculture Adjustment Administration, Soil Conservation Service and Rural Electrification Administration.

relief disbursements. And, as earlier commentators had found, the Committee attributed the distress of the region to overcropping, overgrazing, and the "improper" farm methods being employed. A fundamental cause of the problem, it asserted, was the removal of the region's original grass cover, leading to soil erosion. More specifically the report attributed the heightened vulnerability of the region to drought to a "mistaken homesteading policy and the stimulation of wartime demands." In essence increased cultivation of the region (from 12 million acres in 1879 to over 100 million acres in 1929) enabled by increasingly powerful farm machinery facilitated removal of the grass cover and overgrazing had weakened the remaining cover.

The Committee concluded that, since any permanent increase in rainfall in the region is inconceivable and methods of farming employed by many were better suited to the humid east than to the semiarid Great Plains, farming must be altered to conform to natural conditions. And it further concluded that the farmers of the region were helpless at that stage of the drought and depression to control events; thus a concerted coordinated effort of local, State, and Federal agencies was required to stabilize the region's economy, provide a better, more secure income for each family and spread the shock of inevitable future droughts.

Just as the Roosevelt administration developed agencies and programs to provide assistance and stability to the US Great Plains so did the Canadian government. The Prairie Farm Rehabilitation Administration (PFRA) was established in 1935 in response to the drought and soil erosion occurring in the Prairie Provinces at the time. That agency continues to provide a wide range of financial and educational services to farmers of the region.

5. POST DUST BOWL TRENDS

The long drought ended in the Great Plains and surrounding regions around 1940 as the nation was beginning to emerge from depression—a bit earlier in the Arkansas and the Texas Gulf than in the Missouri Basin (see Figure 2-4). Even before US entry into World War II (December 1941), the war had increased demand for export of agricultural products produced on the Plains. Demand continued to be high in the first postwar years.

Due to the efforts of state and federal researchers yields of the major agricultural crops, including those grown on the Plains, had risen, albeit slowly from the early 1900s to the early 1950s. Thereafter research and development in both the governmental and private sectors contributed to further but more consistent and rapid increases in crop yields. A number of factors have contributed to this phenomenon. Advances in plant breeding permitted the development of crop varieties with increased resistance to disease and insect attack and to greater plant tolerance of temperature and moisture stress. Advances in farm mechanization and the use of chemical fertilizers and pesticides also contributed to rising yields and productivity.

US corn statistics illustrate this phenomenon. About 44 million hectares (109 million acres) of corn were planted in the USA in 1931. Area planted to corn had decreased to 33 million hectares (81.6 million acres) by 2005. Yet from 1931 to 2005 production rose from 56 to 280 million metric tonnes. Yields rose more than sixfold from 1.54 to 9.31 tonne ha^{-1}. Real prices to farmers have been flat or declined over this period.[5] This trend of increasing US corn yields over time is discussed in greater detail in Chapters 4 and 6.

Another major trend has been the growth of irrigation. In 2002 there were about 22.3 million hectares (55 million acres) irrigated in the USA. In 1974 the area irrigated had been 17 million hectares (42 million acres). Since the early 1960s area of irrigated land increased significantly of the Great Plains, most rapidly in the portions overlying the High Plains aquifer (described in Chapter 2). By 1998 Nebraska was second only to California with about 2.4 million hectares (~6 million acres) in area irrigated. The Texas Panhandle and southwestern Kansas had also become major centers for irrigation.

Government support of agriculture did not end with the 1930s drought, as some of its advocates assumed that it would after that crisis had passed. The US Great Plains region continues to get its share—actually more than its share—of dollars under a wide range of federal agricultural support programs.

As is true throughout the USA and Canada, rural population in the Great Plains has declined substantially as a percentage of the total population although in total number that population is more or less stable in the US Plains. And as is also true of both nations, the median age of the rural population and particularly that of farmers has increased from the Dust Bowl days to today. Another noteworthy, if disturbing, fact is that rural counties in the US portion of the Plains are among the poorest in the nation.

It is important in setting the scene to note that the facts and trends enumerated above encapsulate the current condition of the Plains region: aging population, increased production capacity, continued but lessening overtaxing of soil and water resources, fluctuating but often low commodity prices, generally low rural incomes, continuing dependency on governmental support and continuing vulnerability to drought and other climatic stressors. These facts and trends are dealt with in greater detail in the chapters that follow.

6. THE CHAPTERS

As a basis for understanding the region's current problems and future potential Chapter 2 provides a description of the physical environment of the Great Plains region, its boundaries, climate, soils, water resources, original vegetation, and current land use and identifies other regions of the world that are similar in soils, climate, and agricultural potential.

The demographic makeup of the region, its overall economy, and especially its agricultural economy, are described in Chapter 3.

[5] National Corn Growers Association. http://www.NCGA.corn/Ethanol/pdfs/2006/Howmuchethanol can%20comefromcom.v.2.

Chapter 4 has two main aims: (1) to provide a detailed survey of the current geographical distribution of dryland and irrigated crops and animal production on the Plains; and (2) to examine whether or not current uses of the land are "sustainable." Soil erosion and overexploitation of the region's water resources appear to be the most serious threats to sustainability. Therefore, this chapter provides a detailed examination of the current severity of these threats and measures for their control. The concept of a "Buffalo Commons," proposed as an alternative to current use of the Plains region, is examined closely.

Chapter 5 is titled "The Wildcard of Climate Change." Hard as it is to contend with its difficult and highly variable current climate—the factor that most severely limits the long-term productivity and stability of the Great Plains—the prospect that its climate may change (for better or worse) as the result of global warming requires examination. This chapter provides a "mini-primer" on the science of global warming and an assessment of how the Great Plains in particular may be affected. Adaptations to climate change relevant to the Plains are also considered.

Chapter 6 addresses the two key questions raised in this book: (1) Can further global warming (and by inference the threats it poses to the region of interest) be forestalled, controlled, or reversed? (2) Is there a particular role for the Great Plains in any comprehensive strategy aimed at accomplishing this? To address these questions, a detailed review is presented of the role of technology in reducing demand for fossils fuels and in providing substitute supplies of energy. Soil carbon sequestration and biomass for direct combustion or conversion to liquid fuel are the technologies emphasized in this chapter. The scientific background for these technologies, experimental evidence of their efficacy and modeling studies examining their overall potential are described. A separate section of the chapter deals with genetic engineering as it might be applied to increase yields and improve quality of biomass crops, both woody and herbaceous, and as it might help to improve agricultural yields and maintain the capacity to produce food and fiber on a land area reduced in size by a significant conversion of land to biomass production. Economic factors that might favor or restrain the market penetration of soil carbon sequestration and biomass are also examined.

Finally, Chapter 7 summarizes the findings of previous chapters and, in the rapidly changing world of global change science, politics and economics, presents as up-to-date an assessment, as the publication process will allow, of the prospects that there may indeed be "A Biomass Future" in store for the North American Great Plains (NAGP).

REFERENCES

Bonnifield, P., 1979, *The Dust Bowl*, University of New Mexico Press, Albuquerque, 232 pp.

Cooke, M. et al., 1936, Great Plains Drought Area Committee, *Report of the Great Plains Drought Area Committee*, Report to President Franklin D. Roosevelt, The White House, Washington, DC, August 27, 1936. Franklin D. Roosevelt Library, Hopkins Papers, Box 13. http://newdeal.feri.org/Texts/450.htm

Dick, E., 1975, *Conquering the Great American Desert: Nebraska*, Vol. XXVII, Nebraska State Historical Society Publications, Nebraska, 450 pp.

Hurt, R.D., 2004, Dust Bowl, in Wishart, D.J. (ed.), *Encyclopedia of the Great Plains*, University of Nebraska Press, Lincoln, Nebraska, pp. 625–626.

Powell, J.W., 1879, *Report on the Lands of the Arid Region of the United States*, Government Printing Office, Washington, DC.

Smith, H.N., 1950, *Virgin Land: The American West as Symbol and Myth*, Vintage Books (Random House), New York, 305 pp. plus index.

Thompson, J.H., 2004, Settlement Patterns, Canada, in Wishart, D.J. (ed.), *Encyclopedia of the Great Plains*, University of Nebraska Press, Lincoln, Nebraska, pp. 245–246.

Vinton, M.A., 2004, Grasses, in Wishart, D.J. (ed.), *Encyclopedia of the Great Plains*, University of Nebraska Press, Lincoln, Nebraska, pp. 631–632.

Wilhite, D.A., Rosenberg, N.J., and Glantz, M.H., 1986, Improving federal response to drought, *J. Climate and Applied Meteorol.* **25**:332–342.

CHAPTER 2

THE PHYSICAL ENVIRONMENT

1. GEOGRAPHIC BOUNDARIES

We begin with an examination of the NAGP geographic boundaries which have been defined by many, and in many different ways. Depending on whether the boundaries are set by rainfall and altitude or by type of native grassland vegetation (often been defined as the land covered before European settlement by a vast sea of mixed and short grasses) all or parts of the states of Oklahoma, Kansas, Nebraska, North Dakota, South Dakota, and Montana are included. By various definitions, so is much of central Texas, eastern New Mexico, Colorado and Wyoming. Southeastern Alberta, southern Saskatchewan and southwestern Manitoba are included in the more usual map renderings of the NAGP. The extent of scholarly disagreement on this problem of geographic definition is illustrated in Figure 2-1 from the *Encyclopedia of the Great Plains* (Wishart 2004), which represents 50 published versions of the Great Plains regional boundary. Figure 2-2 is the more standard bounding showing the counties included in various analyses throughout this book.

Walter Prescott Webb (1931) set the eastern boundary of the Plains at the 100th meridian; others have moved it to the eastern political boundaries of the states from North Dakota to Oklahoma. And as Figure 2-1 shows, Minnesota, Iowa, Missouri, Arkansas and even Louisiana are granted, by some, the distinction of being a part of the Great Plains. In essence this eastward extension is into what most consider the Prairies rather than the Plains.

The Plains can also be defined as the land lying west of the 20-inch (~500 mm) annual mean rainfall line (isohyet) which coincides closely with the 100th meridian from the South Dakota–Nebraska border down to the Texas–Mexico border. North of Nebraska the 500 mm isohyet moves northeastward, At the North Dakota–Manitoba border it lies close to the 98th meridian. The western boundary of the Plains is less ambiguous, being the foothills of the Rocky Mountains. For the purposes of this book, I take the boundary of the NAGP as that shown in Figure 2-2. Within these defined boundaries the Great Plains encompasses 1,808,170 km^2 or 19.4% of the total area of the USA; the Canadian portion of the Plains occupies 443,289 km^2 or 4.5% of that country's area. Together the US and Canadian Plains cover an area of 2,251,459 km^2.

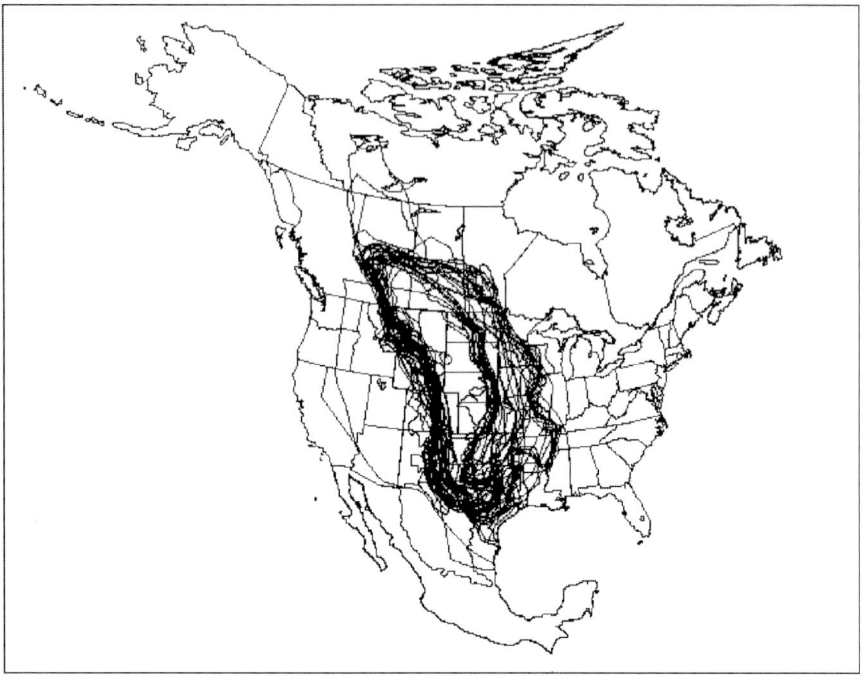

Figure 2-1. Fifty versions of the Great Plains. *Encyclopedia of the Great Plains* (Wishart 2004, p. xvii)

In succeeding sections we explore the climate, soils, water resources and original vegetation of the region bounded by the borders of Figure 2-2.

2. CLIMATE

The Great Plains is a region of extreme and variable climate. A wide range of weather conditions can occur within the period of a day, from one day to the next, from season to season, and from year to year. There are two reasons for this: the greatest portion of the Plains is remote from any major body of water and air masses of differing characteristics alternate frequently in their dominance of the region. Much of the Plains region is remote from the major body of water that influences it most—the Gulf of Mexico. The fact of its remoteness is the reason that the climate is described as continental. Continentality dictates that the diurnal range of temperature (night to day) and the annual range of temperature (winter to summer) will be great. The Great Plains (particularly the northern portion) has the most distinctly continental climate in North America.

Continentality leads to interesting comparisons of the Plains climate with that of other regions of the country. For example, in January the daily minimum temperature in Amarillo, Texas, is the same as that of Detroit, some 800 km

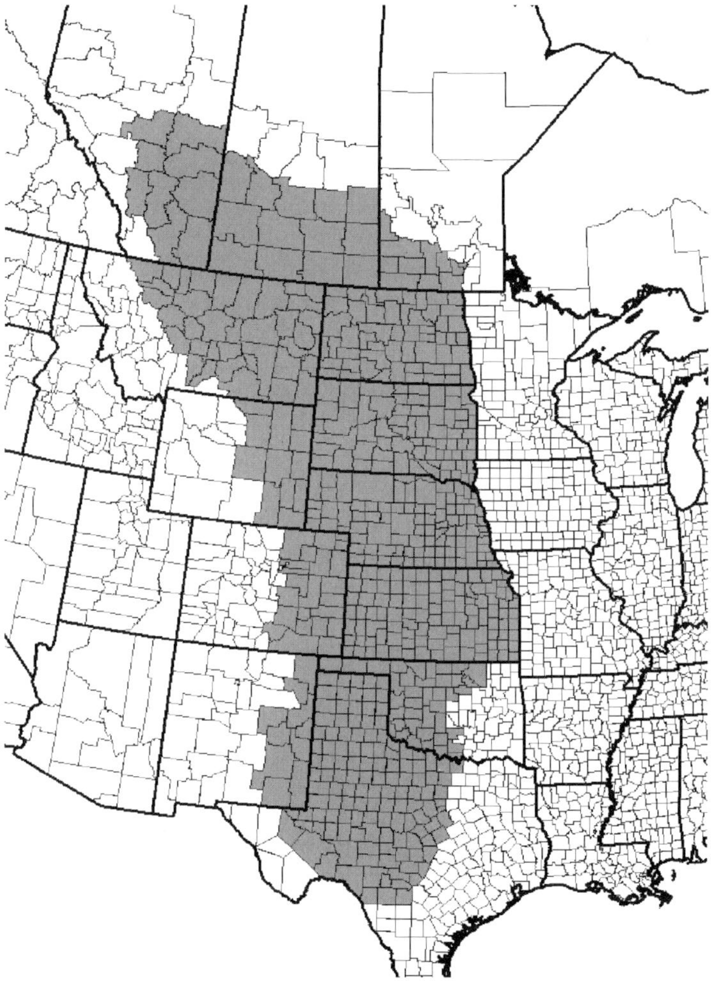

Figure 2-2. US counties and Canadian Census Divisions included in the Great Plains Region

(~500 miles) to its north. In July the daily maximum temperature in north-central South Dakota is the same as that of Jacksonville, Florida. Winters in the northern Great Plains are the coldest in the conterminous 48 states. The southern Great Plains are as hot as the southeastern part of the USA in summer. The northern Great Plains are hotter than the northeastern part of the country. At Steele, North Dakota, for example, the maximum summer temperature recorded is 49.4°C (121°F). The lowest temperature recorded is −27.8°C (−50°F), for an annual range of 77.2°C (171°F). In many parts of the northern Great Plains an annual temperature range of 60°C (140°F) is common.

The Plains region is dominated by the passage of air conditioned over widely different surfaces in regions remote from one another. Air masses conditioned over the tropical seas will be warm and moist; those conditioned over or near the poles will be cold and dry. The region between the Rocky Mountains and the Appalachian chain is open to motion of continental polar air masses that originate in the north and maritime tropical air masses originating over the Gulf of Mexico. The orientation of the North American mountain chains does not separate the air masses from one another as do the Alps in Europe and the Himalayas in Asia. Instead, the air masses meet and clash with great frequency. A third type of air mass enters the Great Plains periodically, that is air conditioned over the Pacific Ocean—moist but cooler than the Gulf air because of its more northerly origin. Maritime Pacific air masses cool as they rise to cross the western mountain ranges, and the capacity of the air to hold water vapor decreases sharply. Condensation and precipitation occur and the air mass dries out. When it descends the eastern slopes of the Rocky Mountains to the western Great Plains, this maritime Pacific air is warm and quite dry. Thus we say that the western Plains lie in the "rainshade" of the Rockies. Still a fourth source of air sometimes dominates the Plains, air warmed over the southwestern deserts of the USA and northern Mexico. Frequent outbursts of warm, dry air in spring and summer from this and other sources create a strong evaporative stress on crops growing in the region.

2.1. Air masses, fronts and precipitation

Rapid changes in the temperature, humidity, cloudiness, windspeed, and wind direction that occur frequently in the Great Plains are due to the passage of air masses with different characteristics: where two or more air masses meet fronts are formed. Cold dry polar air masses moving south encounter warm moist air masses advancing northward from the Gulf. The lighter maritime air is forced aloft and, cooling as it rises, loses water by precipitation. Thus precipitation decreases with distance from the Gulf. From the southeastern US to New England, lines of equal precipitation (isohyets) run more or less east-west. However, the isohyets run north-south from eastern Texas to the northern Prairie states and westward to the Rockies. The drop-off in precipitation with distance from the Gulf is dramatic. Near the Gulf Coast mean annual precipitation is about 1200 mm (~48 inches); and in northwestern Montana it is about 300 mm (~12 inches).

The advance of frontal systems may be accompanied by light general rains or by violent storms in which torrential rains fall for hours or sometimes sporadically for days. The variability in moisture content of the maritime tropical air mass, the path of its movement, and the intensity of its impact with other air masses accounts for the great spatial and temporal variability of precipitation in the Great Plains.

Warming of ground surfaces under unstable moist air may lead to precipitation in isolated thunderstorms. These occur after formation of towering cumulus clouds in summer. The distribution of rainfall from this type of storm is much spottier than from frontal system rainfall events.

The annual patterns of precipitation in the Great Plains change with the seasons. The distribution of snow in the north and rain in the south is fairly uniform from east to west in winter, but winter precipitation contributes only a small portion of the annual total. The bulk of the annual precipitation falls during the growing season from April to September. Autumn weather is usually drier in most of the Great Plains than it is farther east—a definite advantage for the harvest.

It is important to realize that the foregoing describes a normal or average annual pattern of rainfall distribution. Rainfall may be scant during one season and excessive in the next. Drought may occur one year and floods the next. This irregular distribution of rainfall over space and time in the Great Plains is one of the major constraints to the development of the region.

2.2. Evaporation—a critical facet of the Great Plains climate

In the hydrologic cycle the precipitation that falls on terrestrial surfaces has various fates. Some runs off the land into streams and rivers and a portion of this reaches the oceans. Some of the water penetrates the soil and refills the root zone. The excess, if any, seeps down to the underlying water table or until it encounters an impermeable layer. A very large share of the water returns to the atmosphere via evapotranspiration (ET hereafter). ET involves evaporation directly from the upper layer of the soil and transpiration of water drawn through the roots of plants to the leaves where it is also evaporated. ET depends on temperature, humidity, cloudiness and windspeed. For this reason it is a good descriptor of climate overall and its variability in particular. Table 2-1 shows how ET from a standard water-filled pan at a Great Plains location differs from that in other climatic regions of the USA. Coshocton in central Ohio represents the humid Midwest; Davis lies in the Central Valley of California near Sacramento; Phoenix in central Arizona represents a desert climate; Mead lies between Lincoln and Omaha in eastern Nebraska—a region of more moderate climate than most of the Great Plains but nonetheless typical of its hydrologic regime. May to October account for between 75 and 81% of the average annual pan evaporation at all four locations. Interannual variability, represented by the standard deviation of the annual evaporation totals, is greatest at Mead; in fact, double that at any of the other three locations. Evaporation from the Class A pan, while imperfect as a measure of the amount of water that evaporates from lake and reservoir surfaces and definitely different but nonetheless indicative of ET from farmed fields, provides a uniform, standardized way of establishing how the evaporative regime differs from region to region. In the eastern USA and Canada annual precipitation exceeds annual evaporation. In the Great Plains, the opposite holds true. Near Nashville, Tennessee, in the southeastern USA, for example, evaporation from a lake surface will average about 940 mm annually and precipitation will average about 1,220 mm. At North Platte, Nebraska, on the other hand,

Table 2-1. Predicted evaporation from Class A pans and reported maximum ET at four locations in the continental USA (Rosenberg 1969)

Location	Mead, NE	Coshocton, OH	Davis, CA	Phoenix, AZ
Annual pan evaporation (mm)	1,524	1,117	1,778	2,267
Standard deviation (annual) (mm)	264	86	107	173
May–October evaporation (mm)	1,173	861	1,333	1,840
Daily mean May–October (mm)	6.41	4.70	7.28	10.05
Maximum reported daily ET (mm)	12.02	9.14	11.56	12.20
STD/annual pan evaporation (%)	17	8	6	8
May–October/annual pan evaporation (%)	77	77	75	81

the average annual evaporation of 1,170 mm far exceeds the average annual precipitation of 482 mm, a situation typical of the Great Plains and of other arid and semiarid regions around the world.

All of these factors point out that the Great Plains is a region of deficit water supply. The atmospheric demand for water from growing plants is strong and generally exceeds the supply of natural precipitation. It is for this reason that irrigation infrastructure is of such importance in the Great Plains region. Only through the development of irrigation can farmers have a reasonable degree of confidence in their ability to produce consistently high yields of crops and to survive periodic drought.

2.3. Climatic hazards and adaptations on the Plains

The Great Plains posed severe climatic problems for settlers of European origin, problems that to this day restrict the economic and social development of the region. The catalog of climate problems includes extremes of temperature, a growing season limited in portions of the region by low temperatures and/or by dryness, recurrent droughts, strong and persistent winds, and severe storms sometimes accompanied by damaging hail—not to mention tornadoes.

2.3.1. Extreme temperatures

The frequency with which very hot days occur in all parts of the Plains is considerably greater than it is at the same latitudes in the eastern portion of the USA. Forty days a year of temperatures higher than 32°C (90°F) occur at Lincoln, Nebraska. At New York City (about the same latitude) only 5–10 such days occur each year. Great heat imposes a strong evaporative demand on growing crops, which not even irrigation can fully assuage. Human comfort is affected by high temperatures and domestic animals are also sensitive. Large losses of livestock are reported in the Plains each year during hot spells, particularly when humidity is also high.

The polar climate of the Great Plains winter creates other hazards to human and animal life. Good shelter and fuel for heating have always been necessary in the northern Plains. Even in southern Texas, occasional "northers" drop temperatures to below freezing for a few days at a time, damaging sensitive subtropical crops such as grapefruit. Blizzards, although infrequent, create other dangers. People caught out-of-doors can lose all sense of direction during a blizzard and some have frozen to death only yards away from shelter. Herds of cattle drift aimlessly in blizzards and serious losses occur as the animals wander into dangerous terrain.

2.3.2. *Short growing seasons*

Length of the growing season is often defined as the period between the last sustained freezing temperature (0°C; 32°F) in spring and the first in fall. In the northernmost portion of the US Great Plains growing season is limited to between three and four months. In parts of the Prairie Provinces of Canada the freeze-free season may be as short as 60 and as long as 160 days. The choice of crops is necessarily very restricted where growing seasons are short. Settlers on the northern Plains needed short season crops in order to survive and crop research remains devoted to finding species and cultivars (varieties) of crops adapted or adaptable to the short seasons.

In the southern Plains the growing season may last as long as 11 months. In Oklahoma and Texas, subtropical crops like cotton and peanuts are also cultivated along with the standard wheat and corn. Cotton and citrus are grown further to the south. Throughout the Plains region late spring and early fall frosts are possible after planting and before harvest and may cause considerable damage. High value fruit trees are extremely sensitive to frost in the budding stage and many vegetable crops are sensitive throughout their growing seasons. Because of the high frequency of frosts and the considerable cost of frost protection, high value crops are not widely grown north of Texas. The extensive grain and forage cropping systems (large areas, low capital input) that dominate the Great Plains agriculture do not justify the cost of frost protection, so frost is just one more hazard with which the Plains farmers have learned to live.

2.3.3. *Recurrent drought*

Drought can be defined in various ways. Any extended period of dry weather that leads to a measurable loss of crop production can correctly be called drought. Drought can be catastrophic in the Plains region when such periods become so long and the shortage of soil moisture so critical as to cause abandonment of fields already sown, or when crop cover becomes so sparse, because of lack of moisture, to permit the erosion of the soil surface by wind.

There are many definitions of drought, each serving the needs of a particular industry or agency. The National Drought Mitigation Center, based at the University of Nebraska–Lincoln is an authoritative source of information on all aspects of drought. Figure 2-3 from its website[1] identifies three kinds of drought:

[1] www.drought.unl.edu

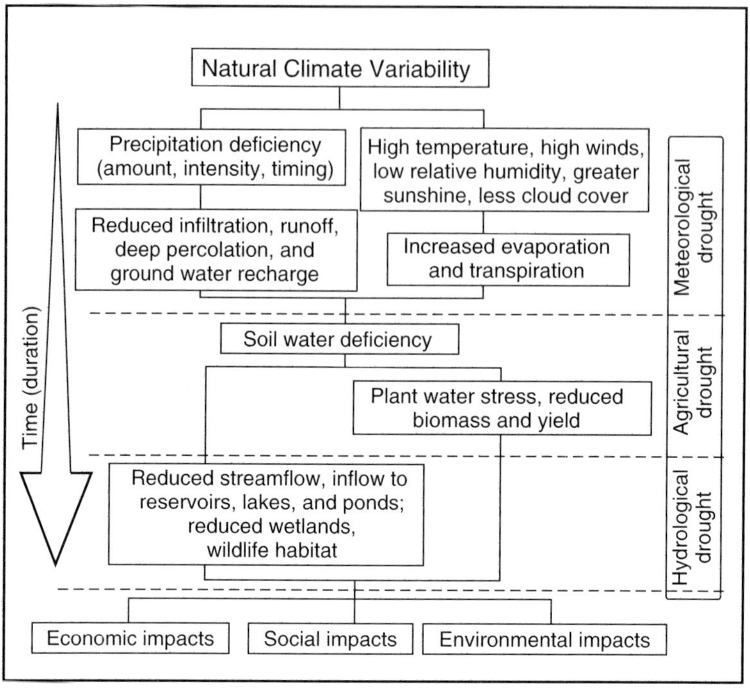

Figure 2-3. Causes and effects of meteorological, agricultural, and hydrological drought. (Source: National Drought Mitigation Center, University of Nebraska, Lincoln, www.drought.unl.edu)

meteorological, agricultural and hydrological. The figure shows how the various meteorological phenomena associated with drought combine to affect agricultural production and how, with increasing duration of the drought, a region's water resource is affected. The figure also shows, importantly, that economic, social and environmental impacts are the end results of severe, protracted drought.

Drought is not new to the Plains. Through dendrochronological studies (tree-ring measurement and interpretation) on specimens of red cedar and ponderosa pine, Weakly (1943, 1963) identified short periods of dry years and, less frequently, droughts lasting for more than five years. Weakly found tree-ring patterns that indicated the occurrence of one drought period of 38 years duration between 1,276 and 1,313 and others of more than 20 years duration.

Long-term records of annual precipitation virtually anywhere in the Plains show extreme year to year variability. Such records also show that dry spells have tended to occur very often in series of 2, 3, or more years—although the longer series are often interspersed with occasional wet years. Thus, for example, in the period since systematic weather records have been kept in the Great Plains, a number of droughts of greater and lesser severity have been experienced.

Through weather records research and the study of newspapers and other historical documents (Palmer 1965) described the drought history of a number of locations in the Plains. The history of western Kansas is typical. A drought that had major impact on the agricultural production in western Kansas occurred in 1894, when as many as 90% of the settlers abandoned their farms in some areas. Drought occurred again in 1913, following the dry years of 1910 and 1911 and the abnormally wet year of 1912. Between August 1932 and October 1940, 34 months of extreme drought occurred in western Kansas. In 1952 and 1953, all 31 counties in western Kansas were declared a drought disaster area by the federal government, and the dry conditions extended through 1955. Similar histories can be recited for other parts of the Great Plains during these same periods. Recurrent drought is part of the Great Plains environment and, unfortunately, its inevitability is often forgotten when the weather is good.

A more panoramic view of drought history in the Plains in modern times (since 1895 in this case) is shown in Figures 2-4a,b,c prepared by the National Drought Mitigation Center (Lincoln, Nebraska). Three of the 18 major river basins in the US—the Missouri, the Arkansas-White-Red, and the Texas-Gulf–cover most of the Plains region. The percentage of land area experiencing severe drought, as indicated by the Palmer Drought Severity Index (PDSI, Palmer 1965) in each year from 1895 to 2004 is shown for each of these basins in these figures. The clustering of drought years is evident in the Missouri basin, with notable and historically important events in the late 19th century, most of the 1930s, the mid-1950s, late 1970s and latter half of the 1980s, and on into the late 1990s and the early years of this century. In the Arkansas-White-Red the 1930s droughts were less severe than in the Missouri basin; in the mid-1950s, however, they were more intense, if shorter-lived. This basin was more strongly impacted in the mid-1960s but less so thereafter. In the Texas-Gulf basin severe droughts appear to recur more cyclically and are more uniform in the extent of land affected.

Another display of information based on the Palmer Drought Severity Index (PDSI) as shown in Figs. 2-5a,b,c in terms of the percent of time that each Climatic Division of the US experienced severe drought (defined by PDSI values of −3.00 to −3.99) during each of three major droughts The northern Plains and Mountain states were the most hard-hit during the drought of the 1930s; the drought of the mid-1950s was most severe in the southern Plains and extended further to the east. The Plains states of Kansas and Nebraska suffered severely in both of these droughts. The worst effects of the drought in 1988 were in the northern Plains states extending almost to the Pacific coast. The central and southern Plains were barely affected.

2.3.4. *Strong and persistent wind*

The Great Plains is the windiest region of the country, open as it is to the free sweep of air masses from the north and the south. Because the terrain is relatively flat and smooth and the land is largely cropped or in grasses, the frictional forces that reduce surface wind speed in other regions are smaller in the Great Plains.

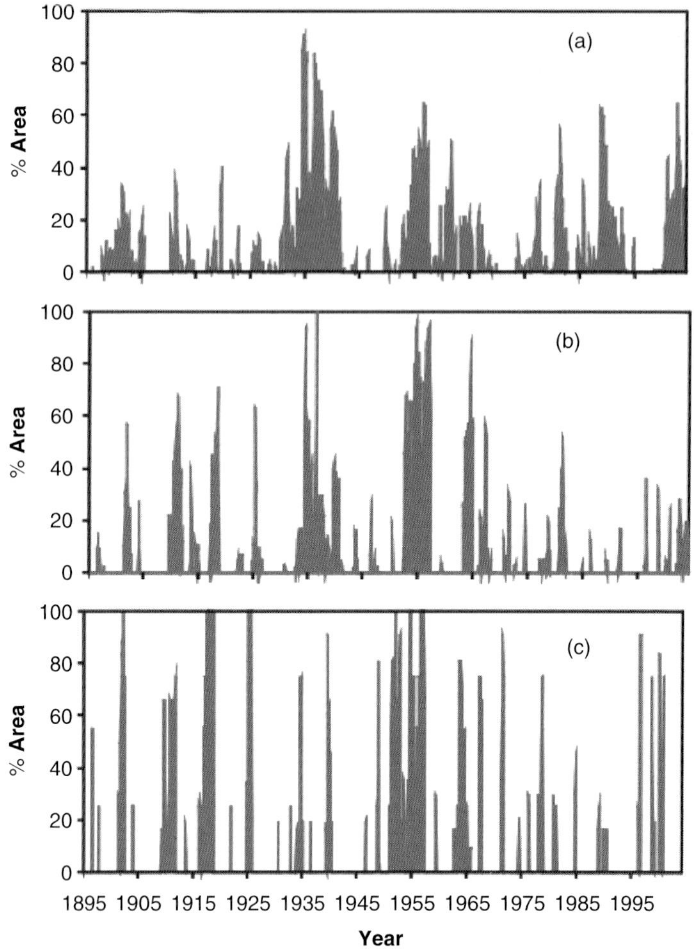

Figure 2-4. Percent area of three Great Plains Basins experiencing severe to extreme drought in the period January, 1895 to March, 2004. (**a**) Missouri; (**b**) Arkansas-White-Red; and (**c**) Texas–Gulf. (Source: National Drought Mitigation Center, Lincoln, Nebraska, based on data from the National Climatic Data Center, National Oceanographic and Atmospheric Administration, US Department of Commerce.)

In winter, northerly winds predominate as far south as Texas. In summer, southerly winds predominate as far north as Montana. Hot southerly winds during the growing season create stress on plants and the low humidity of southwest winds increases the evaporative demand on crops still more.

Strong winds from any direction can cause mechanical damage to crops such as stalk breakage or blowing over (lodging). In winter, northerly winds increase the loss of heat from buildings. Animals also seek shelter from the strong northerly

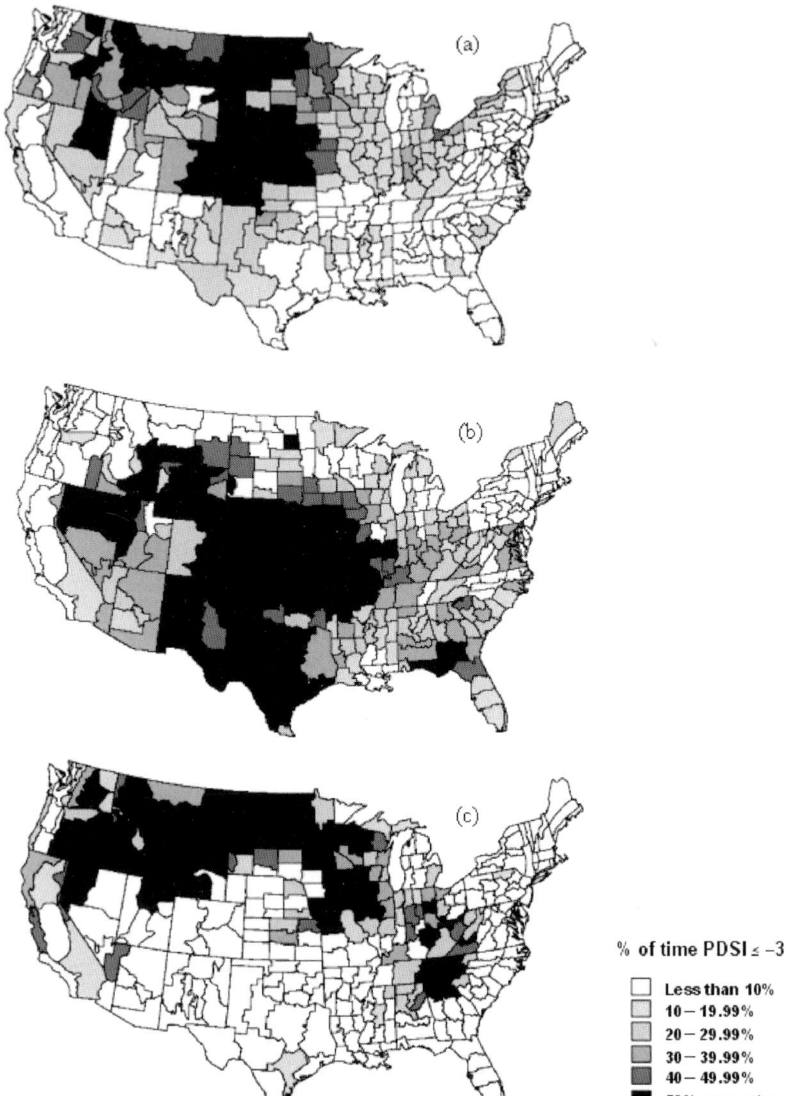

Figure 2-5. Palmer Drought Severity Indices (percent of time in severe and extreme drought) for three droughts in the USA (**a**) 1934–1939; (**b**) 1954–1956; and (**c**) 1988 (Maps prepared from various sources by the National Drought Mitigation Center, University of Nebraska, Lincoln.) (*See Color Plates*)

winds during winter. For these reasons, early settlers used European techniques of windbreak plantations to protect their farmsteads, animals, and fields. In the 1930s and early 1940s, the Shelterbelt Project was initiated by the federal government to protect the entire Great Plains region by the planting of thousands of miles of tree windbreaks. It was thought at the time that, as well as providing immediate protection for fields and farmsteads, this extensive network of trees might beneficially change the climate of the Plains. The Shelterbelt Project was abandoned during World War II for lack of funds and labor to complete it, although windbreak plantings continue less systematically today.

Wind erosion has been a serious problem on the Plains, probably from times well before settlement. During the drought of the 1930s, particularly, farming methods that left the soil bare at certain times of the year led to serious wind erosion and dust storms. The periodic plowing of marginal lands in the semiarid belts of the western Great Plains also contributed to the problem. Modern methods of soil erosion control (strip-cropping, stubble-mulching, tillage to increase surface roughness and more recently minimum-tillage and no-till) have been effective in minimizing wind erosion damage, but these improved practices are not used universally. Wind erosion is dealt with in greater detail in Chapter 4.

The Great Plains settlers saw the wind mostly as an adversary but the wind is, of course, also a source of power. Before rural electrification, windmills were used to drive simple generators. Windmills to pump water for livestock on the range still dot the Plains. With energy shortage before us and with engineering improvements in windmill design, a windmill renaissance may be coming on the Plains. More on this in Chapter 6.

Wind is also the sweeper of the atmosphere. The Great Plains owes the relative cleanliness of its air to the fact that man-made and natural pollutants are rapidly swept out of the region by the almost continual action of winds.

2.3.5. *Other hazards*

Frontal storms over the Great Plains as well as other regions of the country frequently trigger damaging winds and, occasionally, tornadoes. The region of most intense tornadic activity is centered in east-central Kansas and north-central Oklahoma. The frequency of hailstorms is greatest in an area centered in the southwestern corner of the Nebraska Panhandle and in adjacent parts of Wyoming and Colorado. Nine days of hailstorms occur each year, on average. The frequency of hailstorm days drops to about 5 per year in the Wyoming, Kansas, Colorado, and Nebraska areas surrounding the "hail center." The eastern edge of the Great Plains commonly has three to four days of hail storms each year. The high frequency of hail in the western Plains region explains the fact that hail insurance rates in that region are the highest in the USA. Hail can be particularly devastating to wheat growers in the western High Plains because the storms are common during late spring and early summer when the crop is ripening and the hailstones can shatter heads of wheat, making harvesting virtually impossible. Sugar beet cultivation is one means of adaptation to the great hail hazard in western Nebraska, eastern Colorado, and Wyoming. Sugar beets recover remarkably wellfrom hailstorms which sometimes strip the plant of virtually all its leaves. Sugar beet production occurs in the upper Great Plains (north-central Wyoming, Montana, and western North Dakota) and central Grain Plains including southeastern Wyoming, Colorado, and Nebraska.

2.4. Advantages of the Great Plains climate

While it is true that drought, severe weather, strong winds, and extreme heat and cold impose serious restrictions on the developmental possibilities of the Great Plains region, it is also true that the region enjoys distinct climatic advantages. The intensity of solar radiation at any given latitude increases across the USA from the East Coast to the Rocky Mountains. This trend occurs most clearly across the Plains region because of the increasing elevation of the land and the reduced frequency of cloud cover. Because of generally low moisture content the air in the Great Plains region is also less turbid. Thus, crops that require intense sunlight for development are well adapted to the western Great Plains.

Another advantage of the Great Plains climate is the relatively dry autumn season. Over most of the Plains the probability of significant rainfall drops off rapidly in September and October. This makes the use of mechanized harvesting equipment more efficient than it generally is in the more humid east. There are times, of course, when rain and early snowfall make harvesting very difficult, but generally the harvest progresses easily.

Humidity decreases from the east to the west across the USA to the Great Plains and decreases with distance from the Gulf. The low humidity is not conducive to many types of plant fungal diseases that are common in the east. The potato crop in western Nebraska, for example, is more easily protected against "late-blight" (the fungal disease that devastated Ireland's crop in the mid-19th century) than it is in more easterly locations.

2.5. Overview

The climate of the Great Plains is characterized by: a great range in daily, seasonal, and annual temperature; strong atmospheric potential for evaporation because of the ample solar radiation; strong windiness and usual dryness of the air; wide difference in the annual totals of precipitation received from the east to the west, and frequent severe weather including damaging winds, hailstorms, and tornadoes. Limited length of the growing season and irregularity of the precipitation are the major constraints to the stability of agricultural production in the region. Late spring and early fall frosts are frequent but unpredictable. There is a significant risk in the central and northern regions that crops will not have a long enough season for optimum growth. Droughts of greater or lesser severity are a regular feature of the plains climate. The history of the region and its flow of population in and out are closely linked to the incidence of drought, although less so now than in the past.

Adaptation to the Plains climate has required ingenuity, persistence, and fortitude on the part of the peoples who have ventured into it. Earthen shelters and sod houses for protection from the cold; shelterbelts for protection of homes, animals, and fields from the damaging effects of severe wind; use of windmills for power to pump water; introduction of short-season-crop cultivars or those that extend the season by overwintering; soil terracing, stubble-mulching and minimum tillage practices to minimize water and soil erosion; the introduction of irrigation—all are important adaptations that have been made.

Optimists in the past have proposed that man's works would alter the Plains climate—but rainfall does not "follow the plow" as Dr. Samuel Aughey of the University of Nebraska had proposed in the early 1870s, nor, except on the very local scale, do shelterbelts modify and moderate the Plains climate as President Franklin D. Roosevelt's Committee on the Future of the Great Plains suggested in 1936 (Cooke et al. 1936). Neither does it seem reasonable to expect that the influence of large-scale irrigation projects will extend very much beyond the irrigated region. Some believe that cloud seeding for rainfall augmentation will have a significant effect on the climate of the region. Others (the author included) find little evidence to support this belief. Rational planning for the region must recognize that the past climate of the Plains is the best indicator of what it is likely to be at least in the near future. Deliberate amelioration at the hand of man is unlikely. Anthropogenically forced global warming could, of course, alter the climate of the Great Plains. Whether for better or worse is dealt with in Chapter 5.

3. SOILS

In 1941, Hans Jenny, a soil scientist, identified five factors that determine the process of soil formation. These are: nature of the parent material—the rocks or sediments from which the soil is formed; the topography; the climate; the changing vegetation that takes root as the soil forms; the time available to the process. Like all others, the soils formed on of the Great Plains are the product of these factors.

The parent materials from which the soils of the Great Plains have formed were laid down primarily during the Quaternary Era beginning about 1.6 million years ago and much of it during the Era of the Holocene beginning about 10,000 years ago. Swinehart, in the *Encyclopedia of the Great Plains* (Wishart 2004, pp. 629–630), divides the region into three major sectors and lists parent materials as follows: Northern sector covering Alberta, Saskatchewan, Manitoba, Montana, and North Dakota—alluvial deposits, windblown sands and glacial deposits; Central sector covering Wyoming, South Dakota, Nebraska and northern Colorado—all these in addition to windblown silt (known as *loess*); and the Southern sector covering Kansas, Oklahoma, North Mexico, Texas, and southern Colorado—alluvium, lake deposits, windblown sand and silt.

The topography of the Plains, especially in loessial regions is generally less dramatic than in regions to the east and north where receding glaciers laid down deposits in moraines and drumlins. Nonetheless, not all of the region is flat and slopes are sufficiently steep to allow soil to be eroded by water running off the land.

Variations in temperature and precipitation regime determine the organic matter content of the top soil. Organic matter imparts a dark color and contributes to the soil's fertility and physical condition. More organic matter accumulates in the cooler, more northerly soils than in the warmer, southerly ones. Because the eastern portions of the Plains receive more rainfall, soils there have a higher organic matter content than do the dryer soils to the west. Precipitation regime also determines the rate at which calcareous materials are leached from the upper layers and the depth at which they are deposited, sometimes accumulating in indurated layers largely impenetrable to plant roots. The calcareous layer is found at greater depth in the more humid easterly region and nearest the surface in the more arid westerly portions of the Plains.

In a soil taxonomy used by the US Department of Agriculture before 1960 large portions of the Great Plains' most productive soils, were labeled *Chernozem* because of their similarity to the Black Soils of the Ukraine and the more westerly drier soils were simply termed *Brown*. The classification system introduced in 1960 and employed since then by agencies of the US Department of Agriculture, uses a complicated set of descriptive terms drawn from Latin and Greek roots to classify soils[2].

Figure 2-6 shows the distribution of the dominant soil orders in the US and Canadian portions of the Great Plains. In order of area covered these are: *Mollisols, Alfisols, Entisols, Aridisols and Inceptisols*. The Natural Resources Conservation Service (NRDC) describes these soils as follows[3]:

- *Mollisols* are soils that have a dark colored surface horizon relatively high in content of organic matter. These soils are base rich throughout (calcium, magnesium and other metallic ions adsorbed on the surfaces of soil and organic matter particles) and therefore are quite fertile. Mollisols characteristically form under grass in climates that have a moderate to pronounced seasonal moisture deficit. They are extensive soils on the steppes of Europe, Asia, North America, and South America. Mollisols make up about 7% of the world's ice-free land surface;
- *Alfisols* are found in semiarid to moist areas. These soils result from weathering processes that leach clay minerals and other constituents out of the surface layer and into the subsoil, where they can hold and supply moisture and nutrients to plants. They formed primarily under forest or mixed vegetative cover and are productive for most crops. Alfisols make up about 10% of the world's ice-free land surface;
- *Entisols* are soils that show little or no evidence of pedogenic horizon development. Entisols occur in regions of recently deposited parent materials or in areas where erosion or deposition rates are faster than the rate of soil development;

---

[2] Bret Wallach, writing in the Encyclopedia of the Great Plains (Wishart, ed., 2004, p. 617) expresses unhappiness, not unlike my own, with regard to this system, to wit: "In the United States the soil taxonomy used by Marbut was replaced in 1960 by another system, one that introduced an entire lexicon of neologisms and which is therefore exceedingly difficult for all but experts to use comfortably." I recall with sympathy how my Professor of Soil Genesis at the time that this new system was introduced, J.C.F. Tedrow of Rutgers University, disliked it intensely. Nonetheless, we must use it here.

[3] http://soils.usda.gov/technical/classification/orders

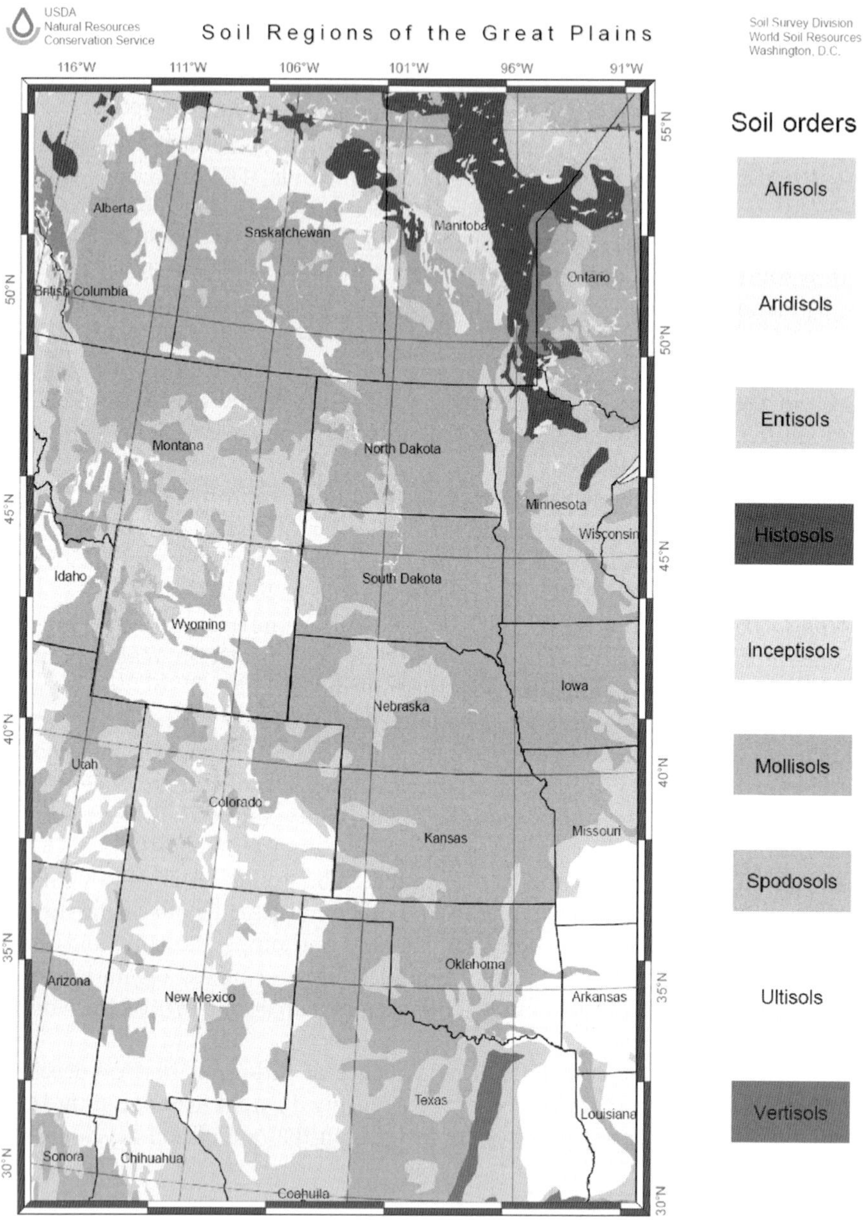

Figure 2-6. Soil regions of the North American Great Plains (Courtesy of US Department of Agriculture, Natural Resources Conservation Service, Soil Survey Division, World Soil Resources, Washington, DC.) (*See Color Plates*)

such as dunes, steep slopes, and flood plains. Entisols make up about 16% of the world's ice-free land surface;

- *Inceptisols* are soils of semiarid to humid environments that generally exhibit only moderate degrees of soil weathering and development. Inceptisols have a wide range in characteristics and occur in a wide variety of climates. Inceptisols. make up about 17% of the world's ice-free land surface;
- *Aridisols* are soils that are too dry for the growth of mesophytic plants (plants living in a temperate environment and receiving average amounts of moisture). The lack of moisture greatly restricts the intensity of weathering processes and limits most soil development processes to the upper part of the soils. Aridisols often accumulate gypsum, salt, calcium carbonate, and other materials that are easily leached from soils in more humid environments. Aridisols are common in the deserts of the world. Aridisols make up about 12% of the world's ice-free land surface.

Figure 2-6 shows that Mollisols dominate the Great Plains region covering most of the southern portions of Manitoba, Saskatchewan, and Alberta, most of North Dakota and Kansas, much of central Oklahoma and Texas, northeastern New Mexico, the northeastern corner of Colorado and southeastern corner of Wyoming, the eastern half of South Dakota, and all of Nebraska except for the Sandhills region in the north-central portion of that state. The Mollisols supported a grass cover at some time in the past; now they are used mainly as cropland. Small grains (wheat, barley) and sorghum are grown in the drier regions; corn and soybeans are grown in the warmer, more humid zones. A mollisol profile is shown in Figure 2-7.

Alfisols are scattered through central Texas and Oklahoma and the Panhandle regions of these states, as well as in northeastern New Mexico and southeastern Colorado with a scattering of these soils in western South Dakota as well. In the Great Plains most of these soils support cropping or grazing.

Entisols are prominent in a cluster on the eastern plains of Montana, Wyoming and Colorado, western South Dakota, and north to south-central Nebraska. These soils are generally used as range or pasture.

Aridisols on the Great Plains are confined mostly to east-central Wyoming and the foothills of the Colorado Rockies. These soils support range vegetation and wildlife habitat.

Inceptisols are found in two large clusters: one is in eastern Montana; the other in west central Oklahoma and adjacent areas in northwest Texas. These lands are used for cropping and pasture.

4. WATER RESOURCES

4.1. Introduction

The Great Plains is a region whose climate varies from the arid in northeastern New Mexico to the sub-humid at the eastern borders of Nebraska and Kansas and central Oklahoma and Texas. So it is a region of deficit precipitation, very

Figure 2-7. A typical Mollisol showing its dark colored surface horizon relatively high in content of organic matter (Source: US Department of Agriculture, Natural Resources Conservation Service, http://soils.usda.gov/technical/classification/orders/mollisols.html) (*See Color Plates*)

much dependant on its water resources for maintenance of its economy. Its rivers and groundwater resources are described below.

4.2. The river basins

The NAGP lies within a number of the Major Water Resource Regions (MWRRs) of the conterminous USA. In the USA the largest part lies within the Missouri MWRR which also drains a small portion of southern Alberta and the Arkansas-White-Red river basins. The Great Plains portions of New Mexico and Texas lie within the Rio Grande and Texas-Gulf MWRRs. Northeastern North Dakota and parts of the Canadian Prairie Provinces are in the Souris-Red-Rainey basin which drains into Canada through Lake Winnipeg and ultimately into Hudson's Bay via the Nelson River. The Saskatchewan River watershed includes much of the Prairie regions of Alberta, Saskatchewan and Manitoba as well as a portion of Montana. It flows eastward emptying in Lake Winnipeg.

Major tributary rivers of the Missouri are the Yellowstone, the Cheyenne, the Platte and the Kansas; of the Arkansas-White-Red, the Canadian; of the Rio Grande, the Conejos, the Pecos and the Alamosa. The Brazos is the major Texas river draining into the Gulf of Mexico.

A sense of the size and flow of these rivers can be gained from Table 2-2, reprinted from Jordan in the *Encyclopedia of the Great Plains* (Wishart 2004, p. 862) which shows the area drained by the major rivers of the region and their average discharge at selected points along their channels for varying periods of record into the 1990s. The "mighty MO" (Missouri) is clearly the largest water-course of the Great Plains region.

4.3. Groundwater

The High Plains or Ogallala Aquifer underlies a large portion of the Great Plains and is its major source of groundwater. Small portions of southern South Dakota and eastern Wyoming, Colorado, and New Mexico are underlain by the aquifer as is most of Nebraska, much of western Kansas, and the Oklahoma and Texas Panhandles. (Figure 2-8). Reports of the US Geological Survey (e.g., McGuire 2004) indicate that extensive use of ground-water from this aquifer for irrigation began in the 1930s and 1940s and increased rapidly from 1940 to 1980. Thereafter, the area irrigated from the aquifer did not change greatly. In fact, area irrigated from the High Plains aquifer has declined from a peak of 5.63 million hectares (13.9 million acres) in 1997 to 5.14 million hectares (12.7 million acres) in 2002. More than 96% of the water pumped from the aquifer has been used for irrigation.

From predevelopment to 2003 the area weighted average water-level changes in the High Plains aquifer declined by 3.84m (12.6ft) (Table 2-3). In South Dakota water level rose slightly during that period. In Texas, water levels declined by 10.6m (35ft) (Figure 2-8). More than a "mind-boggling" 290km^3 (or 235 million acre feet)

Table 2-2. Drainage area and average discharge of Great Plains rivers at various points along their courses (Jordan 2004, p. 862)

Gauging station name	Drainage area (km^2)	Period of discharge analysis	Average discharge ($M^3 sec^{-1}$)
Missouri River at St. Joseph, Missouri	1,100,000	1958–1996	1,300
Missouri River at Bismarck, North Dakota	480,000	1954–1996	650
Missouri River near Culbertson, Montana	240,000	1958–1995	300
Platte River at Louisville, Nebraska	220,000	1953–1996	200
Yellowstone River near Sydney, Montana	180,000	1967–1996	360
Kansas River at DeSoto, Kansas	150,000	1918–1997	210
South Saskatchewan River at St. Louis, Saskatchewan	150,000	1980–1996	190
Arkansas River at Ralston, Oklahoma	140,000	1977–1995	170
North Saskatchewan River at Prince Albert, Saskatchewan	130,000	1980–1996	230
Red River of North Emerson, Manitoba	100,000	1912–1995	100
Colorado River at Austin, Texas	100,000	1937–1996	55
Red River near Gainesville, Texas	80,000	1937–1997	94
Pecos River near Girvin, Texas	77,000	1939–1996	2
Brazos River near Glen Rose, Texas	67,000	1970–1996	34
Canadian River at Purcell, Oklahoma	67,000	1980–1995	22
Cheyenne River at Cherry Creek, South Dakota	62,000	1961–1994	23

of water had been extracted from the aquifer by 2003, more than half of that in Texas. Texas and Kansas together accounted for 80% of the total withdrawals. It is estimated that 185 km^3 (150 million acre feet) had been withdrawn before 1950, when the major expansion of irrigation began in the region. Figure 2-8 also shows the change in water-level that occurred throughout the aquifer from predevelopment to 2003. The greatest declines, 15 to 46 m (50 to > 150 ft) have occurred in the Texas and Oklahoma Panhandles and in southwestern Kansas. In portions of Nebraska, south of the Platte River and at scattered locations in the east and central Nebraska, water levels have actually risen due to seepage from irrigation canals carrying river water originating in the Rocky Mountains.

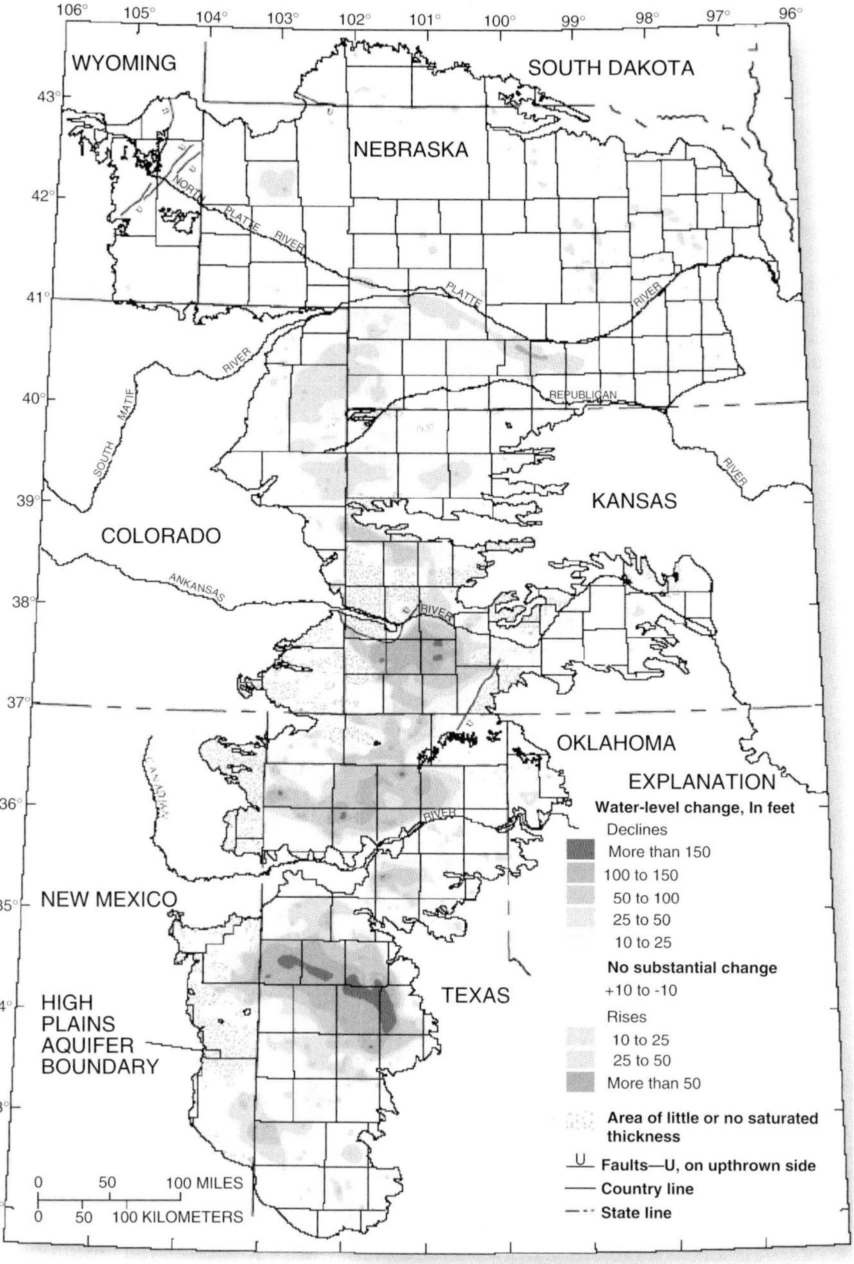

Base from U.S. Geological survey digital data, 1:2,000,000
Albers Equal-Area projection, Horizontal datum NAD83,
Standard parallels 29°30′and 45°30′, central meridian-101°

Figure 2-8. Water-level changes in the High Plains aquifer, predevelopment to 2003 (McGuire 2004, http://pubs.usgs.gov/fs/2004/3097/pdf/fs-2004-3097.pdf) (*See Color Plates*)

Table 2-3. Area-weighted average water-level changes[a] and change in water storage in the high Plains aquifer, predevelopment[b] to 2003 and 2002–2003 (McGuire 2004)

	Area-weighted average water-level change (in meters)		Change in water storage (in km^3)	
State	Predevelopment to 2003	2002–2003	Predevelopment to 2003	2002–2003
Colorado	−3.01	−0.30	−17.1	−1.4
Kansas	−5.75	−0.52	−68.6	−5.1
Nebraska	−0.09	−0.40	−14.1	−10.0
New Mexico	−4.38	−0.18	−11.1	−0.4
Oklahoma	−4.16	−0.24	−14.1	−0.6
South Dakota	+0.06	−0.24	−0.5	−0.5
Texas	−10.61	−0.36	−164.1	−5.2
Wyoming	−0.06	−0.09	−0.6	−0.2
High Plains total	−3.84	−0.37	−290.1	−23.3

[a]Not including areas of little or no saturated thickness.
[b]About 1950.

4.4. Climate sensitivity of the Great Plains MWRRs

Gleick (1990) identified five indicators by which the vulnerability of the MWRRs to climatic conditions might be quantified. These are measures of storage capacity, demand, dependence on hydroelectricity, groundwater vulnerability, and streamflow variability. As given by Gleick:

- The measure of *storage capacity* is defined as the ratio of maximum basin storage volume to total basin annual mean renewable supply. Regions where this ratio is less than 0.6 have small relative storage volumes and, hence, provide little protection from floods and little buffer capacity against shortages. Of the five MWRRs in the Plains only the Arkansas-White-Red is vulnerable on this score although the Texas-Gulf is on the threshold of unacceptability.
- The measure of *demand* is the ratio of basin consumptive depletions (including consumptive water use, water transfers, evaporation, and ground water overdraft) to total basin annual mean renewable supply. Where this ratio exceeds 0.20 water is a decisive factor for economic development. This is the case in the Missouri and Texas-Gulf and especially in the Rio Grande where the ratio exceeds 0.6.
- The measure of *dependence on hydroelectricity* is the ratio of electricity supplied by hydroelectric facilities to the total basin electricity production. More than 25% (0.25) indicates a high dependence on hydroelectricity which, in the case of the Great Plains region, applies only to the Missouri River basin.
- The measure of *groundwater vulnerability* is the ratio of annual groundwater overdraft to total groundwater withdrawals. Regions with ratios greater than 0.25 already have groundwater supply problems. All of the Great Plains MWRRs but the Souris-Red-Rainey are vulnerable on this score.
- Finally, the measure of *streamflow variability* is the ratio of the 5% exceedence flow to the 95% exceedence flow. The former is the streamflow exceeded 5%

of the time—a very high flow. The latter is the streamflow exceeded 95% of the time—a very low flow. Values of 3 and above suggest high streamflow variability. On this score flow variability is high in all of the Great Plains MWRRs.

Thus, the Great Plains basin least vulnerable to climate variability and possible change is the Souris-Red-Rainey although its streamflow variability, especially flooding, is a major problem. The Missouri basin is vulnerable on four counts; only its storage capacity is considered adequate. The other three MWRRs are each vulnerable on three of the measures. The Rio Grande has the highest demand and greatest streamflow variability. The Texas-Gulf MWRR has high streamflow variability and a serious groundwater overdraft problem. The Arkansas-White-Red is low on storage capacity and also has a serious groundwater overdraft problem.

To put the Plains region in perspective it is interesting to note that of the 18 MWRRs of the conterminous US only one—the Great Basin—is vulnerable according to all indicators. California and the Missouri River basin are vulnerable on four indicators; only their storage capacity is deemed adequate.

With regard to storage, it is important to note that the largest reservoir system in North America is that on the Missouri River Main Stem. Its six dams—Ft. Peck (furthest upstream), Garrison, Oahe, Big Bend, Ft. Randall, and Gavins Point (above Sioux City)—can store 90.5 km^3 of water. The system is operated by the US Army Corps of Engineers for the purposes of flood control, navigation, irrigation, hydropower generation, water supply, water quality, recreation, and fish and wildlife (NAS 2002).

Although not fixed by law as priorities, the Missouri River Main Stem System Reservoir Regulation Manual (Master Manual) provides (1) for flood control, (2) for irrigation and other upstream water uses for beneficial consumption purposes, (3) for downstream municipal and industrial water supply and maintenance of water quality, (4) that outflow from Gavins Point provide equitable service to navigation and power, (5) by adjustment of releases from the dams above Gavins Point, power generation to meet the area's needs consistent with other uses and power market conditions, and (6) to the extent possible without interference with the foregoing functions, maximum benefit to recreation, fish, and wildlife.

5. NATIVE ECOSYSTEMS AND VEGETATION

Figure 2-9 divides the conterminous USA into Ecosystem Provinces the nature of which are determined by land-surface form, climate, vegetation, soils and fauna. Four such provinces exist in the region we have defined as the Great Plains and all are within what is called the "Dry Domain". These are the Great Plains Steppe and Shrub Province (311), Southwest Plateau and Plains Dry Steppe and Shrub Province (315), Great Plains-Palouse Dry Steppe Province (331), and the Great Plains Steppe Province (332). The numbers in parentheses identify these provinces in the map and detailed descriptions that follow are drawn from Bailey (1995)[4] and were

---

[4] Bailey, R.G. 1995. Description of the Ecoregions of the United States http://www.fs.fed.us/land/ecosysmgmt/ecoreg1_home.html

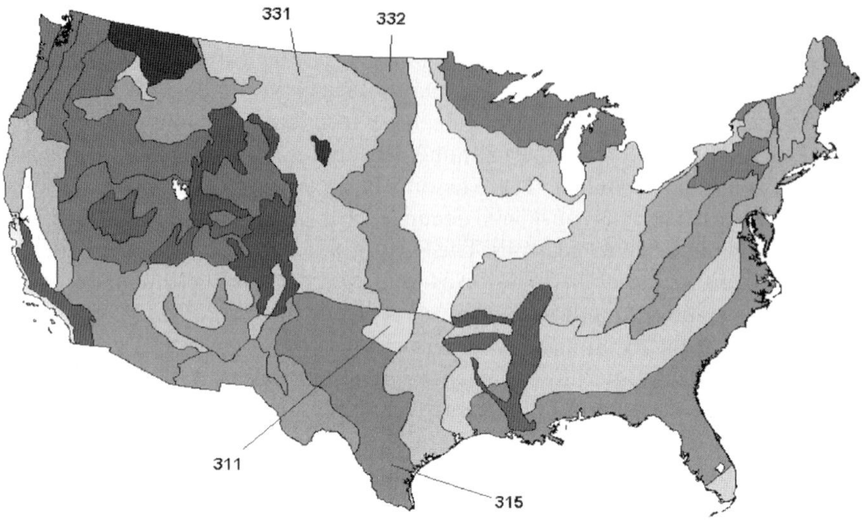

Figure 2-9. Ecosystem Provinces of the Conterminous USA (http://www.fs.fed.us/colorimagemap/ecoreg1_provinces.html)

developed as part of the US Forest Service's National Hierarchical Framework of Ecological Units (ECOMAP 1993). As described in these works:

- The *Great Plains Steppe and Shrub Province (#311)* is found in Oklahoma and occupies 45,600 km^2. Its climate is semiarid-subtropical. Irregular plains dominate the region with elevation gradually increasing from 490 m in the east to 900 m in the west. Soils are Mollisols with some Alfisols. Winters are cold and dry; summers are warm to hot. Average annual temperatures range from 14 to 18 C. Frost-free season ranges from 185 to 230 days. Precipitation ranges from 490 to 740 mm, most falling as rain. Vegetation changes from oak savannah on the eastern boundary to tallgrass prairie of bluestem-grama on finer textured soils through most of the region, changing to sandsage-bluestem on the coarser textures soils near the province's western edge.
- The *Southwest Plateau and Plains Dry Steppe and Shrub Province (#315)* covers much of west Texas and eastern New Mexico. The landform is flat to rolling plains and plateaus. The Stake Plains of Texas and the Edwards Plateau range from sea level to 1,100 m. Altitude of 1,980 m is reached near the Rocky Mountain Piedmont. Climate of this province is semiarid. The frost-free season lasts from only 130 days to more than 300 days. Precipitation falls mostly during the growing season and decreases from 770 mm in the east to 255–380 mm in the west. Evaporation greatly exceeds precipitation in this region. Vegetation in this province is dominated by arid grasslands on which shrubs and low trees grow singly or in bunches. Xerophytic blue grama and buffalo grasses dominate the plains of northwestern Texas and eastern New Mexico but mesquite grows in open stands among the grasses. Oak and juniper are mixed with

grasses, especially needle grass, and mesquite on the Edwards Plateau. Soils in the region include Entisols and support mesquite-live oak savannah; Mollisols associated with mesquite–buffalo grass and juniper-oak savannah.

- The *Great Plains-Palouse Dry Steppe Province (#331)* includes the rolling plains and tablelands sloping down from the foot of the Rocky Mountains and the Missouri Basin Broken lands. Altitude ranges from about 1520m at the foothills of the Rockies to about 760 m at the eastern extreme in the central States. This ecoprovince is mostly flat but punctuated with occasional valleys, canyons and buttes as well, its northern reaches by badlands and isolated mountains. As explained in 2.3.2. the province lies in the "rainshade" of the Rocky Mountains; hence it is dry in the west with increasing moisture from west to east and from north to south because of moisture-laden air from the Gulf of Mexico. Winters are cold and dry; summers warm to hot. Frost-free season ranges from less than 100 days in the north to more than 200 in Oklahoma. The dominant vegetation of this province is shortgrass prairie, typified by bunched and sparsely distributed grasses among which are distributed scattered trees and bushes, among them sagebrush and rabbitbrush. Buffalo grass is prominent. Other grasses include grama, wheatgrass and needlegrass. Although soil organic matter (humus) is in low concentration, overall carbon content may be high because of the presence of precipitated calcium carbonate. These soils, typically Mollisols, are rich in bases. A view of the shortgrass prairie is shown in Figure 2-10.

Figure 2-10. A view of shortgrass prairie near Ft. Collins, Colorado. (Source: Long Term Ecological Research Network, http://savanna.lternet.edu/gallery/sgs/SGS_010016_1) (*See Color Plates*)

- The *Great Plains Steppe Province (#332)* is the narrow strip running from the prairie parkland on its east to about the 104th meridian from Canada to Oklahoma. The land is flat and rolling and well-drained. It rises from 300 m at its eastern edge to 760 m near its west. North of the Missouri River the soils are derived from glacial materials; south of the river from loessial and sand deposits. Precipitation increases from the north to the south and from the west to the east. At the eastern edge of Nebraska precipitation is near 770 mm annually; at its northwestern edge precipitation is close to 380 mm. Soils are mostly Mollisols. Entisols predominate in the Nebraska Sandhills. This mixed-grass steppe transitions from tall grasses prairie parkland at the eastern edge to shortgrass steppe at about the 104th meridian. Tall grasses such as little bluestem and needle and thread grass grow to 1.25 m; the short grasses such as blue grama, hairy grama and buffalo grass to half a meter. Trees are rare except for cottonwood along the watercourses. A number of forbs grow in this ecoregion from Canada to Oklahoma: match weed or broomweed, scurf-pea, sunflower, goldenrod, and ragweed.

The description of the Great Plains ecoprovinces given above, describes the vegetation of uncultivated lands of the region as they were originally. Where land has been left for wildlife and domestic animals to graze, it clearly resembles the pre-settlement ecosystem more than the cultivated lands do. But it would be wrong to assume that the vegetation of the grazed lands still closely resembles what was originally there.

6. LAND USE

Looking at land use in the US Great Plains states in 2002, interesting differences emerge. The states, of course, differ considerably in their total area ranging from Oklahoma, the smallest with 17.8 million hectares to Texas, the largest, in the "lower 48"- with nearly 70 million hectares (Table 2-4).

Colorado had the largest percentage of its total land area in forests (32.2%), with Montana and North Mexico both over 20% forested. By comparison with these mountain states Kansas, Nebraska, and the Dakotas are virtually treeless.

The 2002 US Census of Agriculture indicates that there were 329,000 farms in the US portion of the Great Plains occupying about 155.5 million hectares. Land area in crops is on the order of nearly 68 million hectares. North Dakota is the most heavily farmed of the Great Plains States with 64% of its area in crops, followed by Kansas, Nebraska, and South Dakota.

Pasture and rangeland predominate in North Mexico, Texas and Oklahoma with more than 60% of the land in these three states in pasture or range. The mountain states of Montana and Wyoming and the Plains states of Nebraska and South Dakota are each more than 40% in range and pasture land (Table 2-4).

In the US portion of the Plains, then, pasture and rangeland account for roughly half, cropland for roughly one quarter, forests for about 15% and other uses—urban, rural transportation, rural parks, defense, industrial, and other special uses—account for about 10%.

Table 2-4. Land use in the Great Plains States in millions of hectares. (Source: USDA/Economic Research Service, http://www.ers.usda.gov/data/majorlanduses/2002)

State	Total	Area in millions of hectares				Percentages by land use			
		Forested	Crops	Pasture/range	Other[a]	Forested	Crops	Pasture/range	Other[a]
Colorado	26.96	8.67	4.74	11.85	1.71	32.2	17.6	44.0	6.3
Kansas	21.31	0.9	10.73	7.3	2.38	4.2	50.4	34.3	11.2
Montana	38.08	9.11	7.08	20.4	1.5	23.9	18.6	53.6	3.9
Nebraska	20.03	0.38	9.07	8.32	2.27	1.9	45.3	41.5	11.3
New Mexico	31.4	6.75	1.08	20.9	2.69	21.5	3.4	66.6	8.6
North Dakota	17.87	0.27	11.44	4.86	1.29	1.5	64.0	27.2	7.2
Oklahoma	17.81	3.1	3.19	10.91	0.61	17.4	17.9	61.3	3.4
South Dakota	20.17	0.82	6.76	9.67	2.91	4.1	33.5	47.9	14.4
Texas	69.15	7.43	13.0	43.28	5.45	10.7	18.8	62.6	7.9
Wyoming	25.15	4.45	0.88	11.51	8.31	17.7	3.5	45.8	33
Great Plains	287.93	41.88	67.97	149	29.12	14.5	23.6	51.7	10.1

[a]Includes urban, rural transportation, rural parks, defense, industrial, and other special uses.

The 2001 Canadian Census of Agriculture[5] reports 83,500 farms occupying 39.2 million hectares in our Plains-defined region. Total cropland was on the order of 21.5 million hectares. Range and pasture lands occupied about 4 million and 9.5 million hectares for tame or seeded and natural land pasture, respectively.

7. ANALOGOUS REGIONS

While focused on the NAGP and limited by that region's particular physical and social circumstances, it is possible that our examination (to follow) of that region's potential for biomass production may, if not soon, eventually be relevant to other parts of the world that share some of its characteristics. Which are the analogous regions, if any, for the NAGP? The concept of "agroclimatic analogues", proposed by Nuttonson as far back as 1947, is helpful in identifying such regions and "...in determining which crops will or will not adapt in certain areas and understanding disease and pest susceptibility." Nuttonson's work (1955, 1965, 1966) which provided detailed climatological comparisons of growing regions for wheat and other major crops remains authoritative today.

Wheat is grown widely throughout the world. In one study of climatic requirements for wheat production Nuttonson (1966, Table 1, p. 9) found that *year-round* climates of the Northern Plains from Langdon, North Dakota at 48°55′N to Dodge City, Kansas at 37°45′N are closely mimicked in their annual, warmest month and coldest month maximum, minimum, and mean temperatures, in annual relative humidity and in annual precipitation and its seasonal distributions at locations ranging from western Siberia, central Asia to the Lower Volga region of Russia, the Ukraine, and Turkey. The weather stations used in identifying these analogues covered latitudes from 37 to 52 °N.

Comparing the climates of the *spring-crop season*, Nuttonson (1966, Table 2, p. 10) found agroclimatic analogues for the Ukraine, the Central Chernozem region and the Northern Caucasus of the former Soviet Union as well as Central Asia in the region of Montana, North Dakota, and South Dakota.

By means of a third analytical concept—*year-round global thermal analogues*—Nuttonson (1966, Table 3, p. 11) identified regions that are alike in daylength and temperature conditions but may differ in precipitation patterns and amounts and also in relative humidity. Such differences affect soil moisture conditions but, since deficiencies in soil moisture can be compensated by applications of irrigation water, these analogues are a useful guide for irrigated areas. For portions of the Northern Plains ranging from 48°34′ to 38°30′N and including portions of Montana, North Dakota, South Dakota, Wyoming, Nebraska, and Colorado, thermal analogues were found widely distributed around the Northern Hemisphere: the Northern Caucuses; Sinkiang, China; Vladivostok; southern Manchuria; the Lower Volga; the Ukraine and Northern Japan (Hokkaido).

[5] http://www.statcan.ca/english/freepub/95F0301XIE/tables.htm

Thus it appears that, at least in the Northern Hemisphere, there are many regions that resemble the Great Plains in climate, so that agronomic practices related to biomass production may be transferable among them. The area of potential biomass production expands considerably where irrigation is possible.

8. SUMMARY

At the time of European settlement, the Great Plains region of North America, encompassing more than two and a quarter million square kilometers, was covered primarily with grasses and shrubs—tall grasses in the eastern reaches grading, with declining annual precipitation, through intermediate to short grasses in the west. Today, except for its towns and cities and for small forested zones, almost all the land in this region is used to produce crops or is grazed.

The Plains region is endowed for the most part with deep, productive, mildly alkaline soils. Mollisols predominate and are largely used for farming. Entisols, most in the northwestern portion of the Plains, are largely used for grazing. The inherent productivity of these soils can be reduced by wind and water erosion.

The Plains region is drained by major rivers from the Saskatchewan and Souris-Red Rainey in the north to the Missouri, Arkansas-White-Red, Rio Grande and a number of rivers in the Texas-Gulf region. These rivers provide irrigation water although other uses—navigation, municipal and industrial, fisheries and wildlife–may take precedence depending on season, year, and location. Much of the irrigation water in the Plains is derived from underground sources of which the High Plains (Ogallala) aquifer is the largest. Water withdrawals exceed natural recharge from much of the High Plains aquifer. The problem is most severe in Kansas, the Panhandles of Oklahoma and Texas and the eastern portions of Colorado and New Mexico.

Water that runs off the land to streams and rivers can carry with it sediments, nutrients and chemical pollutants that cause quality problems downstream. Groundwater, too, can be polluted by the leaching downward of fertilizers and chemicals through the soil.

While, as noted above, the soils and water resources of the Plains are subject to degradation, these resources are relatively stable and reliable. It is its highly variable climate that most closely correlates with the region's long-term productivity and which determines more strongly than any other natural factor, its interannual productivity, and economic stability.

The region is classified climatically as semiarid in the western reaches transitioning to sub-humid in the east. Because of its remoteness from the oceans, the climate of the Plains is strongly continental with wide ranges and extremes in daily, seasonal, and annual temperature. Evaporative demand is high but, as is typical in steppe regions, interannual variability in precipitation is great. Drought is a common and recurrent feature of the climate. Tornadoes, hailstorms, and damaging winds are frequent in this region. Winds keep the air relatively unpolluted except when wind erosion in the region or to its west lofts great quantities of soil particles into the air. The Plains are sunnier and less cloudy and humid than regions to the east; the harvest season is generally drier.

While unique in many ways, the NAGP shares certain features with other steppe regions around the world. The Ukraine and other portions of the former Soviet Union and the Pampas of Argentina are examples of "climatic analogues" to the NAGP. Thus, practices for stabilizing agricultural production and enabling biomass cropping in the Plains region, to be examined in subsequent chapters, may be relevant and transferable beyond its borders.

REFERENCES

Cooke, M. et al., 1936, *Report of the Great Plains Drought Area Committee*. Report to President Franklin D. Roosevelt, The White House, Washington, DC, August 27, 1936. Franklin D. Roosevelt Library, Hopkins Papers, Box 13.

Gleick, P.H., 1990, Vulnerability of water systems, in Waggoner, P.E. (ed.), chap. 10. *Climate Change and US Water Resources*, Wiley-Interscience, New York, pp. 223–240.

Jenny, 1941, *Factors of Soil Formation: A System of Quantitative Pedology*, McGraw-Hill, New York, 281 pp.

Jordan, P.R., 2004, Rivers, in Wishart, D.J. (ed.), *Encyclopedia of the Great Plains*, University of Nebraska Press, Lincoln, Nebraska, pp. 861–862.

McGuire, V.L., 2004, *Water-Level Changes in the High Plains Aquifer, Predevelopment to 2003 and 2002–2003*, US Department of the Interior, US Geological Survey Fact Sheet FS-078-03, September 2003.

NAS (National Academy of Sciences) Water Science and Technology Board, 2002, *The Missouri River Ecosystem: Exploring the Prospects for Recovery*, National Academies Press,Washington, DC, 188 pp.

Nuttonson, M.Y., 1955, *Wheat Climate Relationships and the Use of Phenology in Ascertaining the Thermal and Photo-thermal Requirements of Wheat*, American Institute of Crop Ecology, Washington, DC.

Nuttonson, M.Y., 1965, *Global Agroclimatic Analogues for the Northern Great Plains Region of the United States and an Outline of its Physiography, Climate, and Farm Crops*, American Institute of Crop Ecology, Washington, DC.

Nuttonson, M.Y., 1966, *Preliminary Study of Phenological Temperature Requirements of a Few Varieties of Wheat Grown in the Northern Great Plains Region and in Some Agro-climatically Analogous Areas in the Eastern Hemisphere*, American Institute of Crop Ecology, Washington, DC in cooperation with the US Weather Bureau under MU 66-1.

Palmer, W.C., 1965, Meteorological Drought, US Dept. of Commerce, Weather Bureau Research Paper 45, 58 pp.

Swinewhart, J.B., 2004, Geology, in Wishart, D.J. (ed.), *Encyclopedia of the Great Plains*, University of Nebraska Press, Lincoln, Nebraska, pp. 629–630.

Weakly, H.E., 1943, A tree ring record of precipitation in western Nebraska, *J. of Forestry* **41**: 816–819.

Weakly, H.E., 1963, History of drought in Nebraska, *J. Soil and Water Conserv.* **17**:271–273, 275.

Webb, W.P., 1931, *The Great Plains*. Grosset and Dunlap, New York.

Wishart, D.J. (ed.), 2004, *Encyclopedia of the Great Plains,* University of Nebraska Press, Lincoln, Nebraska, 919 pp.

CHAPTER 3

PEOPLE AND THE ECONOMY

1. DEMOGRAPHICS

1.1. Introduction

It is generally well known that agricultural restructuring has dramatically redistributed population in the Great Plains. An analysis by Rathge and Highman (1998) shows that the region's few counties with large urban centers have grown, while the majority of counties, mostly rural, have declined. Prolonged outmigration of young families has distorted the age distribution in many counties and further perpetuated population loss by creating high proportions of elderly. In this section we examine these trends in detail. The point being that, while in population terms the Plains are holding their own and more or less paralleling nationwide trends, rural populations are declining in number and are aging. Many of the statistics presented here are for the Great Plains states rather than for the Great Plains counties as delineated in Figure 2-2.

1.2. Population

1.2.1. Total, rural, and nonrural

From 1900 to 2000 the population of the Great Plains states was in the range of 11–14% of the total US population (Table 3–1). In the 20th century the Plains states population grew from 8.166 to 37.615 million, while the nation as a whole grew from 75.994 to 281.422 million. The Plains states population was therefore ~4.6 times greater at the end than at the beginning of the century while the US population grew by ~3.7-fold.

In the 1990s the US Great Plains states added 6.031 million people; the USA as a whole added 32.712 million. Thus 18% of the US growth occurred in the Plains states during that decade. The US Census Bureau projects that by 2030 the Plains states population will grow by another 15.243 million while the country as a whole grows by 82.163 million. If these projections hold true the Plains

Table 3-1. Rural and total population (in millions) of the US Great Plains states and the USA as a whole, 1900–2000. (Source: US Census Bureau, Decennial Censuses. USA, Regions, Divisions, and States. Table 1: Urban and Rural Population: 1900–1990. Released October 1995. 2000 Census: SFI, American Fact Finder, Table P1)

Year	Great Plains total state population	Great Plains rural population	Great Plains percent rural	US total population	US rural population	Great Plains percent of US total	Great Plains States rural percent of US total rural
1900	8.2	6.5	80.1	76.0	46.0	10.7	14.2
1910	11.2	8.4	74.9	92.0	50.2	12.2	16.8
1920	13.1	9.1	69.4	105.7	51.8	12.4	17.5
1930	15.1	9.5	63.2	122.8	54.0	12.3	17.6
1940	15.6	9.2	58.9	131.7	57.5	11.9	16.0
1950	17.3	7.8	44.9	150.7	54.5	11.5	14.3
1960	20.5	6.8	33.3	179.3	54.1	11.4	12.7
1970	23.0	7.4	32.0	203.2	53.6	11.3	13.8
1980	28.0	7.4	26.5	226.5	59.5	12.4	12.5
1990	31.6	7.8	24.8	248.7	61.7	12.7	12.7
2000	37.6	8.5	22.6	281.4	59.1	13.4	14.4

states will account for ~19% of the nation's growth in the first three decades of this century.[1]

As shown in Table 3-1, the overall rise in population of the Great Plains states has been accompanied by a sharp decline in the proportion of rural people—from 80% in 1900 to only ~23% in 2000.[2] The US Great Plains counties had 141,000 fewer people employed on farms in 2003 than in 1973, amounting to a 25% reduction in the number of workers.[3] The rates of loss were 8.3%, 15%, and 4% in the first, second, and third decades of this period, respectively. The aforementioned trends are also evident in Figure 3-1.

In 1991 the population of the three Canadian Great Plains Provinces was 4.626 million, growing by 9.2% over the decade to 5.054 million in 2001. Alberta led these provinces with growth of 13.1% (Table 3-2).

The US statistics given above are for the Great Plains states. A mid-decadal census reports that between 2000 and 2005, with the exception of gains in western South Dakota, southeastern Nebraska, some of the Texas Panhandle and scattered counties elsewhere in the region, most rural counties in the Plains

[1] U.S. Census Bureau, Population Division, Interim State Population Projections, 2005. Internet release date: April 21, 2005. Table A1: Interim projections of the total population for the United States: April 1, 2000 to July 1, 2030.

[2] The terms "urban", "rural" and "nonrural" have been redefined over time by the US Census Bureau. Consult http://www.census.gov/geo/www/ua/uac2k_90.html for the year 2000 and http://www.census.gov/population/censusdata/urdef.txt for 1900-1990 definitions.

[3] Plains county farm employment data from Bureau of Economic Analysis. 1973–98 Standard Industrial Classification (SIC); 2003 from North American Industry Classification System (NAICS)

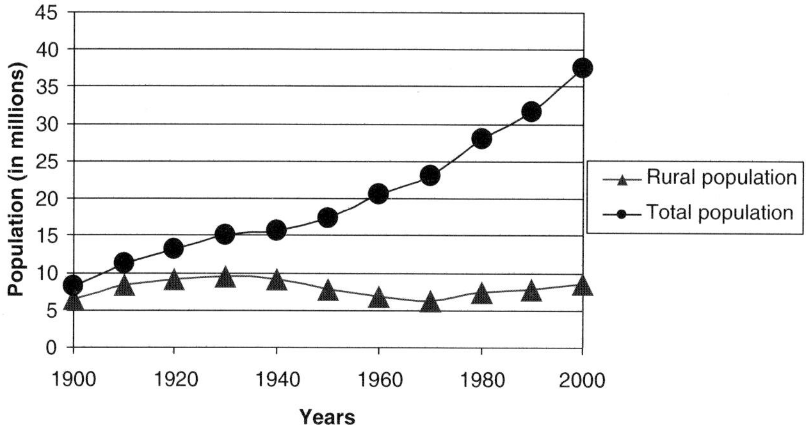

Figure 3-1. Population breakdown of the US Great Plains states. (Source: US Census Bureau, Decennial Censuses. Data at: United States, Regions, Divisions, and States, Table 1: Urban and Rural Population: 1900–1990, Released October 1995. 2000 Census: SF1, American Fact Finder, Table P1).

Table 3-2. Rural and total population (in millions) of the Canadian Great Plains Provinces and Canada as a whole, 1901–2001. (Source: Statistics Canada, Censuses of Population, 1901–2001, http://www40.statcan.ca/101/cst01/demo62a.htm?sdi=population)

Year	Great Plains total province population	Great Plains rural population	Great Plains percent rural	Canada total population	Canada rural population	Great Plains Provinces percent of Canada total	Great Plains Provinces rural percent of Canada total rural
1901	0.4	0.3	75.4	5.4	3.4	7.7	9.3
1911	1.3	0.9	64.7	7.2	3.9	18.4	21.8
1921	2.0	1.3	64.0	8.8	4.4	22.2	28.2
1931	2.4	1.5	62.4	10.4	4.8	22.7	30.6
1941	2.4	1.5	61.9	11.5	5.3	21.0	28.5
1951	2.5	1.4	55.2	14.0	5.4	18.2	26.1
1961	3.2	1.3	42.4	18.2	5.5	17.4	24.4
1971	3.5	1.2	33.0	21.6	5.2	16.4	22.7
1981	4.2	1.2	28.6	24.3	5.9	17.4	20.5
1991	4.6	1.2	25.6	27.3	6.4	16.9	18.5
2001	5.1	1.2	24.3	30.0	6.1	16.9	20.2

continued to lose population, generally between 0 and 1,000 persons. Rural counties gained population in about the same numbers.[4]

A closer look at the rural and urban population distribution of Great Plains counties alone (as of 1996) is given by Rathge and Highman (1998). From 1950

[4] http://www.census.gov/popest/gallery/maps/chg0005.htm

to 1996 the total population of those counties increased from 7.053 to 10.781 million, a total increase of 3.728 million. But the growth in metropolitan areas was 3.950 million, indicating that nonmetropolitan areas lost population. Urban nonmetro counties, defined as counties with a city of at least 20,000 people, also gained population, but rural areas, defined as counties without at least one city of >2,500 people, lost more than half a million people in that period.

Cromartie (1998) has described another relevant trend:

> Over 90% of Great Plains counties experienced an upward trend in net migration from the mid-1980s to the mid-1990s. In that period of time net out-migration continued in sparsely settled, isolated areas and in area where jobs depended on the extraction of energy resources. In-migration during this period was associated mostly with increased commuting from suburban fringe counties or movement to areas high in natural amenities.

1.2.2. Age distribution

At the beginning of the 20th century the average median age of people in the Great Plains states was 22.2 years and of the nation as a whole 22.9 years (Table 3-3). At the end of the century the median ages of these were identical—35.3 years. Through most of the first half of the century the median age of Plainsmen and women was lower than that of the nation as a whole by 1.0–1.6 years. The greatest deviation during the century was noted in the 1960 census when it was 1.8 years. Texas, Oklahoma, and New Mexico had the youngest populations at the beginning of the century when state median age varied from as low as 18.7 (Texas) to as high as 26.6 (Montana). Texas and Colorado were the youngest states in 2000—32.3 and 34.3 years, respectively. New Mexico at 34.6 years is now also among the younger states.[5]

1.2.3. Ethnicity

Most of the people on the US portion of the Plains are of European ancestry. The *Encyclopedia of the Great Plains* (Wishart 2004) provides maps showing the geographic distribution in 2000 census population density by county of Hispanic and Asian origin, African-Americans and Native Americans.

The Hispanic population is most concentrated in the south-central and Panhandle portions of Texas, in eastern New Mexico and Colorado. Hispanics account for more than 80% of the population in some of these counties.

African-Americans are relatively few in the US Great Plains. In a few counties in east-central Texas and several counties clustered at Oklahoma's southwestern border with Texas they constitute 27–28.9% of the population. In scattered

[5] The Census Bureau projects that the US median age will peak at 39.1 in 2035, then decrease to 39.0 by 2050. This is driven largely by the aging of the population born during the "Baby Boom" after World War II. (http://www.census.gov/population/www/pop-profile/natproj.html)

Table 3-3. Median age (in years) in the Great Plains states during the 20th century (see notes below for sources)

State	1900	1910[a]	1920[a]	1930	1940	1950	1960	1970	1980	1990	2000
Colorado	25.9	26.3	27.0	27.3	29.2	29.5	27.9	26.2	28.6	32.5	34.3
Kansas	22.2	23.5	26.0	27.2	30.4	31.1	29.9	28.7	30.1	32.9	35.2
Montana	26.6	26.7	26.9	27.0	28.8	29.9	27.6	27.1	29.0	33.8	37.5
Nebraska	21.6	22.8	25.1	26.3	29.7	31.0	30.2	28.6	29.8	33.0	35.3
New Mexico	21.1	21.3	21.6	21.7	23.0	24.0	22.8	23.9	27.4	31.2	34.6
North Dakota	20.8	21.2	22.1	22.5	25.7	27.1	26.2	26.4	28.3	32.4	36.2
Oklahoma	19.9	20.7	22.2	23.0	26.2	28.9	30.0	29.4	30.2	33.1	35.5
South Dakota	20.7	21.7	23.5	24.4	27.4	28.6	27.7	27.4	28.9	32.5	35.6
Texas	18.7	19.9	22.4	23.7	26.8	27.9	27.0	26.4	28.2	30.7	32.3
Wyoming	24.9	25.2	25.7	26.0	27.6	27.9	27.3	27.2	27.1	32.1	36.2
Great Plains average	22.2	22.9	24.2	24.9	27.5	28.6	27.7	27.1	28.8	32.4	35.3
USA	22.9	24.1	25.3	26.5	29.0	30.2	29.5	28.1	30.0	32.9	35.3

Sources: 1900: Twelfth Census of the USA, 1906. Special Reports: Supplementary Analysis and Derivative Tables. US Government Printing Office. Washington, DC. 1930: Sixteenth Census of the USA, 1940. Population: Volume IV. Characteristics by Age, Marital Status, Relationship, Education and Citizenship. Part 1 US Summary. Government Printing Office. Washington, DC.
[a]estimated

1940	http://www2.census.gov/prod2/statcomp/documents/1951-02.pdf	(p. 36)
1950	http://www2.census.gov/prod2/statcomp/documents/1959-02.pdf	(p. 27)
1960	http://www2.census.gov/prod2/statcomp/documents/1969-02.pdf	(p. 24)
1970	http://www2.census.gov/prod2/decennial/documents/1970a_us1-08.pdf	Table 62
1980	http://www2.census.gov/prod2/decennial/documents/1980a_usC-05.pdf	Table 235
1990	http://www.census.gov/prod/cen1990/cp1/cp-1-1.pdf	Table 251
2000	http://www.census.gov/prod/cen2000/phc-1-1-pt1.pdf	Table 1

metropolitan and nearby counties in these states and in Kansas, Nebraska, and Colorado, they constitute from 9% to 26.9%. In almost all of the remaining counties of the region African-Americans are less than 1% of the population.

Persons of Asian origin are also few in number, constituting no more than 4.5% in a dozen or so counties, mostly near the larger cities. In about 40 of the Plains counties they account for 1–2.9% of the population. In the remainder of this vast region they number under 1%.

Large populations of Native Americans are found in the central and eastern Great Plains counties of Oklahoma, in north-central Nebraska, much of central South Dakota, north-central and western North Dakota, in south-central Montana and along its northern tier of counties. Native Americans constitute more than 25% and often more than 50% of the population in the counties home to reservations and in adjacent counties.

Ethnic population trends for the Great Plains counties and the USA as a whole are shown in Table 3-4. The population of the Great Plains counties grew by 10.3% between 1990 and 2000; the national growth was 13.1%. The Plains are "whiter" than the country as a whole—88% in 1990 and 84.5% in 2000—compared with 80% and 75.1% for the USA in those years. While still a small group on the Plains,

Table 3-4. Population of the Great Plains counties and of the USA as a whole (in millions) by race[a] and ethnicity. (US Census 1990 and 2000[b])

	1990		2000		% Change 1990–2000	
Ethnicity	Great Plains	USA	Great Plains	USA	Great Plains	USA
White	9.73	199.69	10.22	211.46	+5.1	+5.9
African American/Black	0.44	29.99	0.51	34.66	+15.8	+15.6
American Indian/Eskimo	0.22	1.96	0.27	2.48	+21.7	+26.4
Asian	0.12	7.27	0.17	10.64	+47.3	+46.3
Other	0.46	9.81	0.92	22.18	+43.4	+56.6
Total Population	10.97	248.71	12.09	281.42	+10.3	+13.1
Hispanic/Latino origin	0.97	22.35	1.49	35.31	+53.8	+57.9

[a]See http://www.census.gov/prod/2001pubs/c2kbr01-1.pdf for definitions of race categories.
[b]See footnote 5.

the Asian population increased by 47% between 1990 and 2000, little different than the national increase of this ethnic group. In 1990 an "other race" category was used. In 2000 an additional category, "two or more race" was added. The change from "other" to "other plus two or more race" made the largest gain nationally—56.6% between 1990 and 2000. "Hispanic and Latino origin" is an additional category in the census. Between 1990 and 2000 those of this origin increased on the Plains by more than 53%, a few percent less than nationally.

1.2.4. Age of principal farm operators

Trends in the average age of principal farm operators in the Great Plains states and nationally are shown in Figure 3-2. The states of Colorado, Kansas, Montana, and Wyoming show close agreement with the national age of principal farm operators. Oklahoma, New Mexico, and Texas operators are older, while North Dakota, South Dakota, and Nebraska are younger than the national mean. Between 1974 and 1982 the age of operators trended downward by more than 1 year as many "baby boomers" took over from their elders. The age has risen steadily from 50.5 in 1982 to 55.2 in 2002. Virtually all of the Great Plains states show a convergence toward the national average age of farm operators.

1.2.5. Population density

The US and Canadian portions of the Great Plains show different patterns of population density than do these nations as a whole (Table 3-5). The US Plains counties accounted for only 4.3% of the national population in 2000 and 2004. In 2001, the Canadian counterpart area was home to 13.7% of that nation's total population. The US Plains account for almost a fourth of the nation's conterminous land area; the Canadian portion of the Plains accounts for about 7% of land south of the 60th parallel.

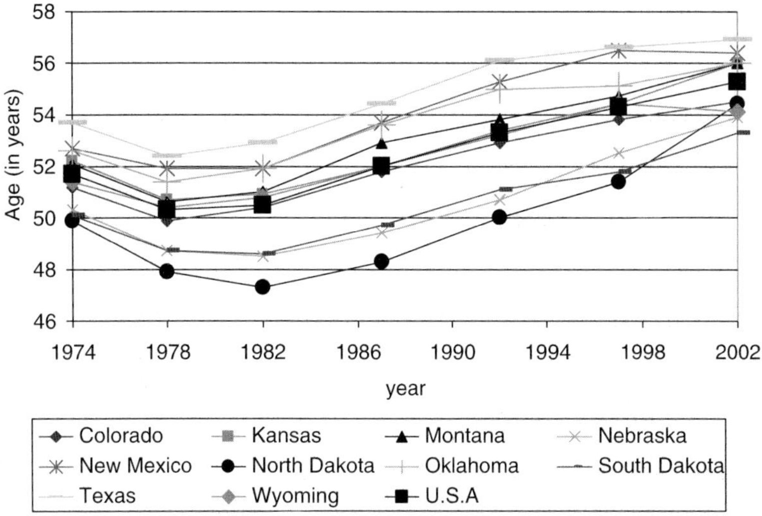

Figure 3-2. Average age of principal farm operators in the Great Plains states and nationally. (Source: 2002 Census of Agriculture. US Department of Agriculture, National Agricultural Statistics Service)

Table 3-5. Population, area, and population density of the Great Plains portions of the USA and Canada and of the nations as a whole. (Source: US (2000) and Canadian (2001) population censuses. 2004 data based on US Census Bureau estimates)

	Units	Country	Year	Great Plains	National	Great Plains % of national
Total population[a]	Millions	USA	2000	12.1	281.4	4.3
		USA	2004	12.6	295.6	4.3
		Canada	2001	4.1	30.0	13.7
Area[b]	$km^2 \times 10^6$	USA		1.1	4.8	23.2
		Canada		0.4	6.0	7.4
Population density[c]	Persons/ Sq. km	USA	2000	10.8	57.8	18.6
		USA	2004	11.2	60.7	18.5
		Canada	2001	9.3	5.0	185.6

[a]USA includes Alaska and Hawaii; Canada includes area north of 60 N latitude.
[b]Area refers to the conterminous US and Canadian Provinces wholly or partially south of 60 N latitude.
[c]Population density refers to the number and density of people only within the indicated area.

Population density in the US Plains grew from 10.8 in 2000 to 11.2 in 2004, remaining at about 18.5% of national population density. In Canada (area south of the 60th parallel only) population density in 2001 was 5 per sq. km; in the Plains region it was significantly higher—9.3 per sq. km.

2. THE GREAT PLAINS ECONOMY

2.1. Great Plains and national gross domestic product

In 2003 the US Great Plains states accounted for $1.33 of the $10.29 trillion, or 12.9%, of the national gross domestic product (GDP), up by 0.2% from 1997 (Table 3-6). While the national GDP rose by 19.4% from 1997 to 2003, the GDP of the Plains states as a whole rose by 21.5%.

Private industries accounted for 87.6% of the Plains economy in 2003, not much different from the national figure of 88.7%. Agriculture (crop and animal production) accounted for only 0.7% of the private sector economy nationally in 2003. The Plains states are twice as dependent on basic agriculture (1.4%). However, food-product manufacturing is a larger factor nationally (1.5%) than in the Plains states (1.2%). Forestry, fishing and related activities contribute only 0.2% to the private sector Plains economy compared with 0.3% nationally. On the national scale forestry, fishing, and related activities are 43% as large as crop and animal production. In the Plains states these activities contribute less than one-fifth of what crop and animal production do to the economy, despite the importance of tourism and recreation in the Rockies.

During the period 1997–2003 notable changes occurred in the relative importance of the individual sectors to the Plains states' economy as a whole. While the region's economy grew by 21.5% as a whole, forestry, fishing, and related activities grew by 54.5%, crop and animal production by 31.2%, the information and finance and insurance industries by 57.4% and 44.4%, respectively. Mining revenues were down by 22.6%.

Major changes in the Plains states' share of the national economy are also represented by sector in Table 3-6 for the period 1997–2003. These are: 3.5% increase with respect to national crop and animal production; 0.4% drop in food product manufacturing; 4.6% drop in the mining sector; 0.9% increase in the information sector and 0.3% increase in the government contribution to the economy.

The large role of the Plains states in mining (52.7% of that sector's national product in 2003) is, perhaps, misleading with respect to the Plains region *per se*, since most mineral extraction (coal, uranium and metals) occurs in the Rocky Mountains and western portions of Wyoming, Colorado, and New Mexico. Oklahoma and Texas account for most of the petroleum and natural gas extracted on the Plains proper. The states wholly within the Plains region—North Dakota, South Dakota, Nebraska, and Kansas—together account for only 2.4% of the mining economy of the Great Plains states. It is difficult to establish just what portion of the mining sector product resides within the Plains counties of the other Plains states. An approximation can be obtained from the proportion of personal income derived from oil and gas in these counties as a fraction of total personal income. In Colorado that fraction is <1%; in New Mexico 7.3%; in Oklahoma 2.1%; in Texas 4.0%; and in Wyoming 3.9%.

The product value of the information and finance and insurance sectors increased greatly on the Plains (54.7% and 44.4%, respectively) and nationally

Table 3-6. US Plains state and national gross domestic product (in thousands of 2000 dollars). (Source: US Department of Commerce, Bureau of Economic Analysis—Regional Economic Accounts, http://www.bea.gov/bea/regional/reis/)

Industry	1997	2000	2003	2004	% Change 1997–2004	2003 % of USA
Total gross state product	1,096,183	1,253,719	1,332,107	1,387,076	26.54	12.9
Private industries	948,330	1,098,784	1,167,527	1,217,833	28.42	12.8
Agriculture, forestry, fishing, and hunting	16,061	17,594	21,523	17,796	10.80	20.8
Crop and animal production (farms)	13,794	14,741	18,126	NA	NA	24.9
Forestry, fishing, and related activities	2,224	2,852	3,397	NA	NA	10.9
Mining	71,215	58,178	55,077	57,351	−19.47	52.7
Manufacturing	134,510	158,592	163,014	171,664	27.62	11.3
Food product manufacturing	16,753	16,191	15,985	NA	NA	10.3
Information	47,047	69,689	73,970	79,104	68.14	14.7
Finance and insurance	60,759	74,772	87,843	93,132	53.28	10.3
Government	148,200	154,936	164,572	169,269	14.22	14.0
Federal civilian	28,142	27,703	27,117	NA	NA	12.3
Federal military	16,429	16,328	18,178	NA	NA	18.5
State and local	103,658	110,907	119,196	NA	NA	13.9

(47% and 44.3%, respectively) in the period 1997–2003. The relative importance of the Plains information sector grew by 0.9%.

Of particular interest here is the fact that crop and animal production are a very small component of both the gross state and gross domestic product. On the Plains, however, that industry's contribution to the total economy is roughly twice that of the country as a whole ($75.97 billion out of a total GDP of $10.24 trillion or 0.7%). Also of special interest is the fact that the Plains states accounted in 2003 for almost exactly one quarter of the total US product in that sector, rising from 1997 by 3.5%. Food product manufacturing is smaller proportionately on the Plains than nationally, its percentage of national product falling slightly from 1997 to 2003.

2.2. Personal income

Total personal income, defined as total active income (earnings), passive income and government transfers, is another good measure of the economic strength or weakness of a county, state, or region. Total personal income in the Great Plains in 2002 was 1.153 trillion dollars (Table 3-7), 12.6% of the US total. Nonfarm income was 12.5% of the national product. Only $11 billion of the Plains total product was farm income, less than 1% of the Plains total. While a small fraction, it is twice that for the nation as a whole. Further, farm income on the Plains is almost exactly one quarter of the national farm income. Farm income is greatest in Texas and Nebraska and least in Montana. Farm income contributes most to the total personal income in the Dakotas.

Another telling indicator of economic well-being is average wages and salaries. A review of data from income tax returns for the tax year 2003 indicates that, once government payments are removed from overall income, 27 of the lowest 50 wage/salary-earning counties are located on the Great Plains. Of these Montana has one, North Dakota, South Dakota, and Nebraska have seven each, Texas has two, and Colorado, Kansas, and New Mexico have one each. Average county salary and wages ranged from $13,485 in Meagher County, Montana, to $17,356 in Hayes County, Nebraska. By way of comparison, the highest county level average wage/salary was $74,416 in Somerset County, New Jersey. Interestingly, another seven of the 50 lowest earning counties are located in the Great Plains states but outside of our defined area of interest.

The Economist magazine[6] comments on this statistic, calling the northern Great Plains "America's new ghetto" and pointing out, as the figures in Table 3-4 confirm, that the population of this "ghetto", except for "several pockets of wretched Native American poverty" is largely white.

---

[6] *The Economist*. December 10, 2005. Not here, surely? The poorest part of America. pp. 31–32. Original data source: Transactional Records Access Clearinghouse, Syracuse University. http://trac.syr.edu/tracirs/findings/aboutTP/

Table 3-7. Total personal income,[a] farm and nonfarm income in the Great Plains states and nationally, 2003. (Source: US Department of Commerce, Bureau of Economic Analysis—Regional Economic Accounts, http://www.bea.gov/bea/regional/reis/)

	Personal income billion ($)	Nonfarm income billion ($)	Farm income billion ($)	Farm % of personal income (state)
Colorado	157.2	156.5	0.7	0.45
Kansas	80.2	79.5	0.8	0.95
Montana	23.3	23.2	0.2	0.72
Nebraska	52.4	50.5	1.9	3.67
New Mexico	47.0	46.3	0.6	1.34
North Dakota	18.3	17.4	0.9	5.04
Oklahoma	93.7	92.5	1.2	1.27
South Dakota	22.1	21.1	1.0	4.48
Texas	642.6	639.1	3.5	0.55
Wyoming	16.3	16.1	0.2	1.19
Great Plains total	1,153.1	1,142.1	11.0	0.95
US total	9,151.7	9,107.5	44.2	0.48
Great Plains % of USA	12.6	12.5	24.9	

[a]As defined by BEA, personal income is the income that is received by all persons from all sources. It is calculated as the sum of wage and salary disbursements, supplements to wages and salaries, proprietors' income with inventory valuation and capital consumption adjustments, rental income of persons with capital consumption adjustment, personal dividend income, personal interest income, and personal current transfer receipts, less contributions for government social insurance.

2.3. The farm economy and government payments

Table 3-8 presents information on the numbers of farms in the Great Plains counties and the land area they occupy. Additionally the table shows the extent of the government support payments distributed to these farms—all in comparison with these statistics for the nation as a whole.

The Plains counties hold over 329,000 land units identified as farms, 15.5% of the US total. These farms are large—a necessity where precipitation is the limiting factor for crop production—accounting for nearly 41% of the total farmed area of the USA But farms in this region are on average more dependent on government support than is true of the US as a whole. The 15% of the nation's farms located in the Plains received 31.5% of the government payments distributed to farms in 2002.[7] About 55% of the Great Plains farms receive payments compared to 33% nationwide. The average per hectare payment to the Plains farms was, however, less than the nationwide average—$13.20 compared to $17.24.

[7] Government payments to farmers include incentives for adoption of conservation measures, commodity price supports, insurance coverage for losses due to natural disasters, supports to farmers for transitioning from tobacco to other crops, and a wide range of additional programs.

Table 3-8. Farms, land area, government payments to farms and farms receiving government payments on the county level. (Source: the US Department of Agriculture—National Agricultural Statistics Service 2002 Census of Agriculture)

State	Number of farms (×1,000)	Land area in farms (ha×1,000)	Number of farms with cropland (×1,000)	Total cropland (ha×1,000)	Number of Farms with harvested cropland (×1,000)	Area of harvested cropland (ha×1,000)	Number of farms with Irrigated Land (×1,000)	Area of irrigated land (ha×1,000)	Number of farms receiving government payments (×1,000)	Government payments ($×1,000)
Colorado	16.1	8,782	10.9	3,696	6.8	1,151	4.8	597	7.6	110,423
Kansas	64.4	19,112	56.7	11,955	44.1	7,679	5.9	1,083	39.2	328,244
Montana	17.5	20,408	14.2	6,648	10.4	3,095	4.3	385	10.5	198,351
Nebraska	49.4	18,576	43.7	9,113	37.1	7,015	18.0	3,085	32.0	347,517
New Mexico	4.7	6,528	3.2	662	1.7	162	1.7	183	2.1	39,180
North Dakota	30.6	15,902	28.5	10,726	20.8	8,056	0.7	82	23.9	293,067
Oklahoma	36.4	8,954	28.4	4,181	19.4	2,245	1.8	182	16.5	128,680
South Dakota	31.7	17,719	28.1	8,222	22.9	5,460	1.8	162	20.3	215,084
Texas	73.3	29,963	53.7	8,791	32.6	3,790	10.8	1,637	26.6	372,404
Wyoming	5.2	9,566	3.5	754	2.3	264	1.8	236	2.1	26,488
Total Great Plain	329	155,514	271	64,752	198	38,923	52	7,637	181	2,059,438
USA	2,129	379,708	1,751	175,700	1,362	122,497	299	22,383	707	6,545,678
Plains % of USA	15.5	41.0	15.5	36.9	14.5	31.8	17.2	34.1	25.5	31.5

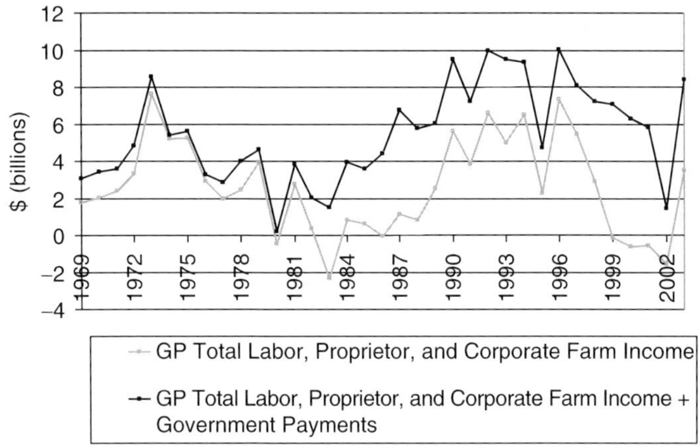

Figure 3-3. Trends in farm income from the sale of crops and livestock at the Great Plains County level (unadjusted for inflation) (Source: US Bureau of Economic Analysis, Local Area Farm Income and Expenses data. Data originally derived from USDA-NASS.)

Texas has the largest number of farms in the Plains region, the largest land area and the smallest percentage of its farms receiving the greatest state total of government payments. The percentage of farms participating in government payment programs is greatest in North Dakota, South Dakota, and Nebraska, closely followed by Kansas. Wyoming and Montana receive the smallest per hectare payments. Reasons for these differences are explored in Chapter 4.

Despite movements in the US Congress in recent years to reduce supports for agricultural commodities, dependency on the government for support to Great Plains farms continues, as is shown in Figure 3-3 which is not adjusted for inflation.

3. SUMMARY

The Great Plains is a sparsely populated region. In the US portion of the region the rural population is about the same now as it was in 1900 but, as is true of the nation as a whole, its percentage of the total population has declined sharply. Rural counties containing metropolitan areas have shown growth, but counties lacking cities greater than 20,000 continue to lose population. Overall, the median age in the Great Plains increased from about 22 to 36 years from the beginning to the end of the 20th century. The median age of principal farm operators in the USA has risen from the late 1970s and early 1980s from about 50 to 55+. Individual Great Plains states bracket the national average by +/−2 years. The population of the Great Plains counties is distinctive in its "whiteness"—85% compared with the national average of 75%. It is also among the poorest regions—by some measures it is the poorest—in the nation.

We picture the Great Plains states as predominantly agricultural (crops and animal grazing). In terms of land use that is certainly the case. But only about 1.4% of the gross regional product, which was about 13% of the gross national product in 2003, derives there from crop and animal production. Yet that surprisingly small percentage actually represents 25% of the national crop and animal product. Only in North Dakota does the farm percentage of state personal income exceed 5% (just barely). And a considerable portion of the farm income in the region comes from government payment; in some years, farm income is positive only because of government payments.

This quick overview of population and economy suggests that the Great Plains is a region in which, at this time, all is not well. What are the prospects for reversing the less positive current demographic and economic trends in the region? A very complicated question, the answers to which will depend on a future of many and perhaps unforeseeable political, economic, technical, and social developments. But one factor, a more sustainable, more profitable agriculture will have to be part of the answer. A broader view of the region's current agriculture and its associated environmental problems is presented in the following chapter.

REFERENCES

Cromartie, J.B., 1998, Net migration in the Great Plains increasingly linked to natural amenities and suburbanization, *Rur. Dev. Perspec.* **13**:27–34.

Rathge, R. and Highman, P., 1998, Population change in the Great Plains: A history of prolonged decline, *Rur. Dev. Perspec.* **13**:19–26.

Wishart, D.J. (ed.), 2004, *Encyclopedia of the Great Plains*, University of Nebraska Press, Lincoln, Nebraska, pp. 5, 137, 350, 560.

CHAPTER 4

AGRICULTURE AND SUSTAINABILITY

1. INTRODUCTION

The nature of contemporary agriculture in the US and Canadian Plains is described in the following sections as are the associated environmental problems of soil erosion and groundwater depletion. An alternative future for the Plains, posits the notion that its current agriculture is unsustainable because of unsuitable use of the land and overuse of its water resources, and that the region must be returned to grass cover. The validity of this prescription is examined in the light of changing management practices and applications of new agricultural technologies.

2. THE AGRICULTURAL ENTERPRISE

2.1. Crop production

Statistics from the 2002 US Census of Agriculture on total cropland, harvested cropland, and land in irrigation were given by state in Table 3–8 for the US Great Plains counties. Only those counties shown in Figure 2-2 are included. The primary land use in the Great Plains is for agriculture—with the largest area devoted to annual crops. In 2002, there were some 329,000 farms on the US portion of the Plains, 15.5% of the national total. Total cropland in the Plains is about 64.8 million hectares which is 36.9% of the US total, indicating that farms are larger than average in the Plains. More than 7.5 of the 64.8 million hectares are irrigated—about one-third of the US total. There are 52,000 Great Plains farms that have irrigation. Nebraska alone accounts for about 40% of the total irrigated acreage; Texas, Kansas, and Colorado account for 21%, 14%, and 8%, respectively. In 2002, 38.9 million hectares of crops were harvested in the US Plains region.

Area planted in 2002 to the major crops grown in the US Plains are listed in Table 4-1. Of these crops wheat occupies the greatest area (~11.8 million hectares in 2002), followed by corn (~7.0 million hectares), soybeans (~5.6 million hectares), sorghum (~2.0 million hectares), and cotton (~1.7 million hectares). Other crops of importance include sunflower seed, barley, dry edible beans, oats, sugar

Table 4-1. Crop area and production on the US Great Plains counties. (Source: US data from the 2002 US Census of Agriculture. World data from the US Department of Agriculture Foreign Agricultural Service)

Crop	Area ha × 1000	Production tonne × 1000	% of USA Area	% of USA Production	% of World Area	% of World Production
Corn	7,045	47,502	23	21	5	8
Wheat	11,758	22,347	64	52	5	4
Oats	269	428	33	25	2	2
Barley	628	1,453	39	31	1	1
Sorghum	2,008	5,192	69	61	5	9
Soybeans	5,645	12,230	19	17	7	7
Dry edible beans[a]	394	739	58	55	–	–
Cotton	1,658	927	33	25	5	4
Potatoes	57	1,607	11	8	–	–
Sunflower seed	694	865	94	93	4	4
Canola[b]	732	488	94	95	2	2
Sugar beet for sugar	160	7,448	29	27	–	–
Peanuts	119	402	24	28	1	1
Vegetables	37	–	2	–	–	–
Orchards	51	–	2	–	–	–
Forage[c]	7,192	31,663	28	20	–	–

[a]Excluding limas.
[b]Data from the North Dakota Oilseed Council.
[c]Land used for all hay, haylage, grass silage, and greenchop.

beets, peanuts, and potatoes. Relatively small areas are devoted to vegetables and orchards production. Wheat is the major crop on the Canadian Plains (Table 4-2) accounting for nearly 10 of the 23.3 million hectares in annual crops. Canola is produced on ~4.6 million hectares followed by barley (~4.1 million hectares), oats (~1.4 million hectares), dry peas (~1.3 million hectares), and flaxseed (~0.7 million hectares). Crops produced in lesser quantities include mustard seed, sunflower seed, rye, and mixed grains.

The geographic distribution of the major crops described above is represented for all of the conterminous USA in Figure 4-1. These data represent the predominant crop, by county, during the period 1985–1997. Continuous winter wheat dominates the eastern two-thirds of Kansas and Oklahoma, some of the Texas Panhandle and eastern New Mexico. Spring wheat dominates in eastern North Dakota. Most of the western and northern Plains region is in a wheat (winter or spring)—fallow rotation. Corn and soybeans in rotation dominate northeastern Kansas and eastern Nebraska and South Dakota. Corn dominates central Nebraska as does cotton in the Texas Panhandle. Table 4-1 for the USA and Table 4-2 for Canada provide details on Great Plains crop production.

Table 4-2. Crop area and production, Canadian Great Plains census divisions.[a] (Source: Canadian data, http://www1.agric.gov.ab.ca/$department/deptdocs.nsf/all/sdd10293. World data from the US Department of Agriculture Foreign Agricultural Service)

Crop	Area ha × 1000	Production tonne × 1000	% of Canada		% of World	
			Area	Production	Area	Production
Corn (all)	123	1,592	8	9	<0.1	<0.1
Wheat (all)	9,934	21,053	95	89	5	4
Oats	1,385	3,174	88	86	10	12
Barley	4,111	11,256	92	91	7	8
Rye	101	212	69	65	–	–
Mixed grains	32	80	24	21	–	–
Canola	4,633	6,566	99	98	17	10
Flaxseed	729	754	100	100	–	–
Dry peas	1,268	2,115	100	100	–	–
Mustard	328	226	100	100	–	–
Lentils	536	520	100	100	–	–
Sunflower seed	115	150	100	100	<0.1	<0.1
Tame hay	4,870	11,204	68	50	–	–

[a]Manitoba and Saskatchewan acreages are for 2003. All other data for 2002 including Alberta areas are for 2002.

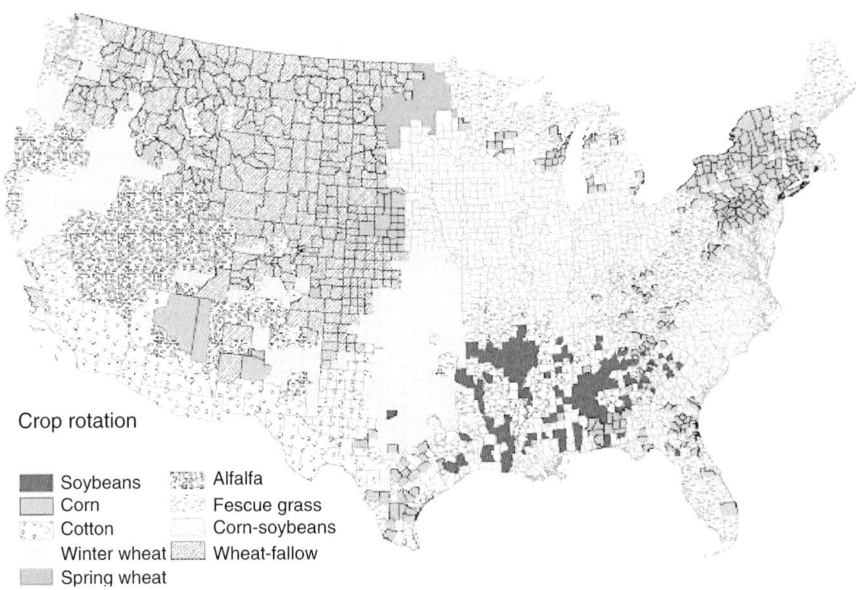

Figure 4-1. The predominant crop by county averaged from the 1985 to 1997 US Natural Resources Inventory. (Courtesy of C. Brosch and R.C. Izaurralde, Joint Global Change Research Institute, College Park, Maryland) (*See Color Plates*)

Wheat: (*Triticum aestivum* L.) Winter wheat accounted for 63%, spring wheat 29%, and durum wheat 7% of the total US wheat production in 2002. About two-thirds of the total area in winter wheat and 83% of the total production occurs in Kansas, Oklahoma, Nebraska, and Colorado. Small amounts of winter wheat are grown in Canada. Spring wheat is the major crop grown in the Prairie Provinces of Canada (Table 4-2). In the USA, spring wheat is confined to the northern tier of states. North Dakota alone has over 2.3 million hectares; about another million hectares are planted in Montana and South Dakota. Together these states produced 99% of the 6.155 million metric tonne 2002 crop. Durum wheat, used primarily for pasta, was grown on nearly 1 million hectares in the USA during the 2002 crop season, predominantly in North Dakota with the remainder in Montana and South Dakota. North Dakota accounted for 80% of the 1.6 million tonne US production.

The US Plains produce 52% of all US wheat on 64% of the land in wheat production (Table 4-1). The Plains account for 47% of the winter wheat, 58% of the spring wheat, and 75% of durum wheat production in the USA. Canada's wheat crop is 76% spring, 20% durum, and only 4% winter. Forty-four percent of Canada's Great Plains 7.2 million tonne spring wheat crop is grown in Saskatchewan, 33% in Alberta, and 23% in Manitoba. Manitoba produces 62% of the Plains 227,000 tonne winter wheat crop; Alberta and Saskatchewan each produce 19%. Seventy-five percent of Canada's 2.5 million tonne durum wheat crop was produced in Saskatchewan and 24% in Alberta.

The US Plains provide 5% of the world's wheat land, yielding 4% of the global production. Canada, too, produces 4% of the world's wheat on 5% of the area devoted to that crop (Tables 4-1 and 4-2). Overall, in the 2001/2002 trade year, Canada and the USA exported about 82% and 49% of their total production (16.8 and 26.3 million tonnes, respectively).

Corn: Corn (maize, *Zea mays* L.) production on the Great Plains is second in area only to wheat. The Canadian Plains does not produce significant quantities of corn. Eighty-nine percent of the US land in corn is farmed to produce grain; the remainder is for silage. The 6.3 million hectares used to produce grain on the US Plains constitute 23% of the US total area and 5% of the total global area devoted to this crop. The 47 million tonnes produced on the US Plains constitutes 21% of the national production and 8% of global production. Nebraska is the largest producer of corn for grain (~23 million tonnes on 3 million hectares), followed by South Dakota (~7.5 million tonnes on 1.3 million hectares) and Kansas (7.4 million tonnes on ~1 million hectares). North Dakota, Colorado, and Texas together produce about 8 million tonnes on slightly less than 1 million hectares. Much of the corn grown on the Plains is irrigated, accounting for the relatively high yields in this semiarid region.

Soybeans: Soybean (*Glycine max* (L.) Merr.) is not an important crop on the Canadian Plains. In 2002 soybeans were grown on ~5.6 million hectares on the US portion of the Plains (19% of the US producing area) yielding ~12.2 million tonnes (17% of national production). The Plains produced 7% of the global crop on 7% of the land devoted to soybeans. The primary producers of soybean are Nebraska, South Dakota, North Dakota, and Kansas with, respectively, 38%, 28%, 20%, and 13% of the region's production. Soybeans are grown primarily in the eastern portions of these four states although plant breeding has facilitated its movement somewhat westward into central Nebraska and Kansas in the last two decades.

Sorghum: Sorghum (*Sorghum bicolor* (L.) Muench) too is unimportant on the Canadian Plains. On the US portion of the Plains about 94% of this heat-loving tropical grass grown is for grain; the remainder is grown for silage. Sixty-one percent of the US production stems from the Plains region where it is grown on 69% of the land allocated to that crop nationally. The Plains yields 9% of the global production on 5% of the sorghum land. The leading producer is Kansas with 61% of the Plains land allocated to sorghum and 62% of the region's total production. Texas has the next largest area and production (22% and 21%, respectively), followed by Nebraska and Oklahoma.

Cotton: Cotton (*Gossipium hirsutum* L.) grows only in the southern reaches of the Great Plains. The region accounts for 25% of the national production on 33% of the land devoted to it nationally. The Plains contribution is 4% of the global crop grown on 5% of the land. Texas dominates cotton production on the Plains—93% of the Plains area and 92% of the 2002 production. Oklahoma, Kansas, and New Mexico also produce cotton. In Texas about half the cotton acreage is fully or partially irrigated.

Canola: Canola (*Brassica napus* L.), as its recently applied name implies, is important in Canada (canola is otherwise known as rapeseed) (Table 4-2). Ninety-nine percent of the Canadian crop stems from the Plains provinces. Saskachewan produces 41% of the 6.6 million tonne crop. Alberta and Manitoba produce 33% and 26%, respectively. The USA as a whole produced only 488,000 tonnes of canola in 2002. North Dakota provided 91% of that crop on 91% of the land devoted to it. Kansas and Montana report some canola production but, for all intents and purposes, the crop belongs to North Dakota.

Sunflower seed: Sunflower (*Helianthus annuus* L.) is particularly important to the agriculture of the Dakotas (Table 4-1). The Plains provide 94% of the US land planted to sunflower and 86% of that is in the Dakotas. Ninety-three percent of the 0.86 million tonne national production stems from the Plains and 90% of that from the Dakotas. North Dakota produces about four times more than South Dakota. The Canadian Plains produce 150,000 tonnes of sunflower seed, 83% of that in Manitoba.

Barley: Barley (*Hordium vulgare* L.) a cool season small grain, grows primarily in the northwestern portion of the US Plains and in the Prairie Provinces. The US Plains produce 31% of the national production on 39% of the area devoted to this crop. Great Plains barley production is only 1% of global production. North Dakota and Montana provide, respectively, 84% and 11% of the land area in barley production on the Plains and essentially the same proportions of the production. The Plains contribution of ~11.3 million tonnes is 91% of the Canadian production. Alberta, Saskatchewan, and Manitoba produce 49%, 39%, and 12% of that crop, respectively. The Canadian Plains produce 8% of the world crop on 7% of the land used for its production.

Oats: Oats (*Avena sativa* L.) are grown in the three Prairie Provinces and all of the US Plains states. The Canadian Plains produce 86% of Canada's 3.2 million tonne crop on 88% of the land planted to that crop. Of that Saskatchewan provides 38%, Manitoba 34%, and Alberta 28%. The Plains region provides 33% of the land devoted to this crop in the USA and produced 25% of its crop. On the Plains the largest area and production stems from North Dakota—45% of the area and 44% of the crop. South Dakota contributes 20% of the area and 21% of the crop. Texas, Nebraska, and Kansas are the next largest contributors.

Dry edible beans: The US Plains is also a major provider of dry edible beans (*Phaseolus vulgaris* L.). Fifty-eight percent of the US land devoted to this crop is found in the region and it is responsible for 55% of the 739,000 tonne national production. North Dakota accounts for 68% of the area and 62% of the Great Plains production. Nebraska and Colorado provide 15% and 7% of the area and 21% and 9% of the production of this crop on the US Plains.

Sugar beets: The Great Plains region also contributes 27% of the 7.45 million tonne national sugar beet (*Beta vulgaris* L.) crop. North Dakota contributes 65% of the Great Plains production grown on 65% of the area. Together Montana, Colorado, and Nebraska produce another 34% of the crop on 32% of the area. Although previously grown in a number of provinces, Alberta was the only remaining producer of sugar beets in Canada in 2004. About 744,000 tonnes were produced on about 14,500 ha.[1]

Potatoes: The Plains produce 8% of the US potato (*Solanum tuberosum* L.) crop. Potatoes are produced in the more northerly Plains states, particularly North Dakota on 75% of the area with 65% of the production and in Nebraska, 16% of the area and 24% of the production. Colorado, Kansas, and Texas are minor producers of this crop. Canada produced 4.3 million tonnes of potatoes in 2005, about 38% from the Prairie Provinces. Of that quantity (1.65 million tonnes) Alberta produced 49%, Manitoba 44%, and Saskatchewan only 7%.[2]

Peanuts: The US Plains also produce 28% of the national peanut (*Arachis hypogaea* L.) crop. Eighty percent of the Plains peanut crop comes from Texas; the remaining 20% from Oklahoma and New Mexico.

2.2. Irrigation

Irrigation is important to the agriculture of the Great Plains region. Water is supplied from the river systems and from aquifers underlying the region. These resources were described in Chapter 2. The irrigated area is large; about 7.78 million

[1] http://www.agr.gc.ca/misb/spec/index_e.php?s1=bet&page=intro

[2] http://dsp-psd.pwgsc.gc.ca/Collection/Statcan/22-008-XIE/22-008-XIE2005003.pdf for potatoes

hectares in 1997. By 2002 that area had decreased to 7.64 million ha. Irrigation, as shown in Figures 4-7 and 4-8, is not uniformly distributed but is concentrated in certain portions of the region. The 2002 US Census of Agriculture shows that the largest cluster is in central and eastern Nebraska where, in many of the counties, more than 40% of the land is irrigated. The Panhandle of Texas, southwestern and south-central Kansas, and eastern Colorado have large areas of land under irrigation. In many of their counties 10–40% of the cropland is irrigated. Irrigation is scattered in the Dakotas; few counties have more than 5% of the land in irrigation. Irrigation is more widespread in eastern Wyoming and Montana with several counties having up to 10% of their land under irrigation.

The major irrigated crops of the region are corn, sorghum, wheat, and cotton. Corn irrigation is concentrated in Nebraska. Significant areas of corn irrigation are also found in northeastern Colorado and the Texas Panhandle. Corn is also irrigated in a few scattered counties in the eastern Dakotas. Sorghum irrigation is concentrated in southwest Kansas and the Oklahoma and Texas Panhandles. Irrigation of winter wheat is concentrated in the Oklahoma and Texas Panhandles with additional significant areas in southwest Nebraska, northwest Colorado, and southwest Kansas. There are also significant areas of spring wheat irrigation in Northern Montana and western North Dakota. Cotton irrigation is most important in the Texas Panhandle and eastern New Mexico. Cotton is irrigated also in the Oklahoma Panhandle and southwest Kansas.

According to the UN Food and Agriculture Organization, the total land area irrigated in Canada in 2001was 785,000 ha. The Prairie Provinces accounted for three-fourths of the total land area irrigated: Alberta, 63%; Saskatchewan, 8.7%, and Manitoba, 3.6%.[3] Sugar beets, vegetables, fruit, oats, alfalfa, and barley are the principal irrigated crops in Alberta.[4]

2.3. Animal production

Table 4-3 shows that cattle and calves produced in the US Plains counties and sold in 2002 numbered over 36.6 million representing nearly 50% of the national production. Of this inventory beef cattle predominate. Texas, Kansas, and Nebraska accounted for 23%, 21%, and 20% of this production, respectively, each moving more than 7 million animals. Colorado and Oklahoma each moved more than 3 million cattle and calves. New Mexico and Wyoming sold less than 1 million cows and calves in 2002. The census divisions on the Canadian Plains produced over 7.7 million cattle and calves in 2001 (Table 4-4), also about half of that nation's production.

The US Plains counties produced and sold over 21 million hogs and pigs in 2002, 11.5% of the national total. Nebraska was by far the largest producer with

[3] http://www.fao.org/AG/AGL/AGLW/aquastat/irrigationmap/irritabcanada.htm

[4] http://www.albertasource.ca/alphabet/article.php?article_id=302

Table 4-3. Numbers of livestock in the US Great Plains counties, by state. (Source: 2002 US Census of Agriculture)

	Cattle and calves sold (× 1000)	Hogs and pigs sold (× 1000)	Sheep and lambs inventory (× 1000)	Broilers and other meat-type chickens sold (× 1000)
Colorado	3,166	1,272	156	7
Kansas	8,044	3,512	81	92
Montana	1,328	239	249	47
Nebraska	7,351	8,994	97	3,361
New Mexico	699	2	54	0.1
North Dakota	1,100	394	114	186
Oklahoma	3,006	3,087	47	0.6
South Dakota	2,708	3,774	377	321
Texas	8,488	56	908	7
Wyoming	763	6	258	0.7
Great Plains total	36,653	21,334	2,340	4,022
US total	73,509	184,998	6,342	8,500,313
Plains % of USA	49.9	11.5	36.9	<0.1

Table 4-4. Livestock on the Canadian Great Plains Provinces, by census division. (2001 Canadian Census of Agriculture 2001[a])

	Total cattle and calves (× 1000)	Total pigs (× 1000)	Total sheep and lambs (× 1000)	Total hens and chickens (× 1000)
Manitoba	799	1,185	59	3,231
Saskatchewan	2,153	736	114	3,903
Alberta	4,780	1,587	198	8,895
Great Plains total	7,732	3,508	371	16,030
Canada	15,551	13,959	1,262	126,160
Plains % of Canada	49.7	25.1	29.4	12.7

[a]http://www.statcan.ca/english/freepub/95F0301XIE/tables/html/Table23Can.htm

42% of the total. South Dakota, Kansas, and Oklahoma follow with 17%, 16%, and 15%, respectively. The Canadian Plains produced about 14 million pigs, 25% of the national total. Alberta was the largest producer.

Sheep and lambs in inventory numbered about 2.3 million in 2002 on the US Plains, about 37% of the national inventory. The largest producers—Texas, South Dakota, Wyoming, and Montana—together account for three-quarters of the region's sheep and lamb production. Texas alone accounts for 39%. The Canadian Plains in 2001 produced 371,000 sheep and lambs, about 30% of the national production. Alberta was the largest producer.

Overall, poultry is not of great importance on the US Plains. Broilers and other meat-type chickens sold numbered just over 4 million out of a national inventory of 8.5 billion birds. Total hens and chickens were four times more numerous on the Canadian Plains—over 16 million. The Canadian Plains contribute about 13% of the national production.

In summary, the Great Plains plays a disproportionately large role in US cattle and sheep production, providing about one-half and one-third of the national supply, respectively. The region produces a small but significant portion of the nation's hog production. Plains poultry production is insignificant on the US national scene.

The Canadian Plains is also a dominant producer of that nation's cattle (~50%) and produces a relatively larger share of its nation's hogs, sheep and poultry than the US Plains do.

2.4. Government payments to agriculture

As was shown in Chapter 3 in which the overall economy of the US Great Plains is described, government payments provide a substantial share of the income to farmers in the region. The 2002 US Census of Agriculture shows that 25% (~181,000) of the 707,000 US farms receiving some governmental support are in the Plains region, but their share is a bit larger than average: $2.059 billion of the $6.546 billion national payout (~31.5%) reaches Great Plains farmers. Eighty percent of the beneficiary farms are located in (from most to least) Kansas, Nebraska, Texas, North Dakota, and South Dakota. These five states receive ~75% of the federal dollar payments to the area. Texas replaces Kansas as the leader in terms of dollars received.

Do these payments make prospects for development of a sustainable agriculture and land management regime more or less likely? Can agriculture in this region survive without government payments? These questions are explored in sections that follow.

3. THE CONCEPT OF SUSTAINABLE AGRICULTURE

3.1. Introduction

The issue of global sustainability gained prominence with the release in 1987 of a report entitled *Our Common Future*, produced under the auspices of the World Commission on Environment and Development and chaired by the then prime minister of Norway, Dr. Gro Harlem Bruntland. The report was, in a sense, a catalog of global problems ranging from the increasing world population, species extinctions, urbanization, energy, and food security problems. With regard to the latter, the Commission emphasized that the problem of increasing food production to keep pace with demand, while retaining the essential ecological integrity of production systems, is serious and difficult but that sustainability can be achieved with new technologies to increase agricultural productivity while conserving land and water resources, if appropriate agricultural policies are adopted.

In the context of the Great Plains (or any other agricultural region) a sustainable agriculture requires that its natural resources, soil and water especially, be used efficiently and conserved over the long term. As Francis (2004) puts it, sustainability "...means an agriculture that does not overly exploit or deplete these resources...is productive, economically sound, environmentally benign and socially viable."

The Great Plains may seem, in the public imagination, to be an unlikely candidate for a sustainable agriculture to take hold. Images of the Plains from the time of European settlement through the great droughts of the 1890s, 1930s, 1950s, and subsequently convey the impression that agriculture on the Plains has been anything but sustainable what with crop failures, soil erosion by wind and water and consequent out-migration of rural populations and collapse of the small towns that provided them service and support. Figures 4-2 and 4-3, classic photos from the drought era of the 1930s, vividly portray the "Dust Bowl" experience. The questions of whether sustainability is a possibility for the Plains and, if so, whether biomass cropping can contribute to that goal are explored below.

3.2. Recipes for sustainable land use on the Plains

The generally pessimistic view of the Plains as unsustainably managed has prompted thinking on alternative ways for managing the land resources of the region. The work of Popper and Popper (1987, 1999) is a recent and imaginative contribution to this thinking. They have argued for a return of much of the

Figure 4-2. "Fleeing a dust storm." Cimarron County, Oklahoma. Arthur Rothstein, photographer, April 1936. (Library of Congress)

Figure 4-3. Prowers Co., Colorado, 1937. (Western History Collection, University of Oklahoma)

sparsely populated portions of the Great Plains land to grassland. The area so restored would become a "Buffalo Commons" stretching from Mexico to Canada through which buffalo will again migrate freely.

The Great Plains as delineated by the Poppers extends from the foothills of the Rocky Mountains to the 98th meridian, more or less the regional boundaries used in this book. In their 1987 paper the authors state:

> We believe that over the next generation the Plains will, as the result of the largest, longest-running agricultural and environmental miscalculation in American history, become almost totally depopulated.

They offer a plan for a phased "deprivatization" of the land, and its transfer back to governmental control and management.

In the late 1980s, when the Buffalo Commons concept was put forth, economic times on the Plains were difficult. The Poppers point to the emptying of small towns, the aging of the populace, soil erosion "approaching Dust Bowl rates" and impending water shortages, not least from the mining of the Ogallala aquifer. At that time, too, energy prices were low, depressing income in areas of the Plains whose economies depend on oil and natural gas extraction. The "Boom and Bust" economy of the Plains region was viewed by the Poppers as heading inevitably to the latter condition—a permanent bust, an emptying out of the High Plains, making a return to what Licht (1997) had termed "a more compatible and environmentally friendly use"—a "Buffalo Commons" encompassing 360,000 km^2 (139,000 sq. miles) of wildlife refuges. In such a case, tourism would provide economic opportunities now lacking. In the years since the Poppers' first paper on the Buffalo Commons the Plains region has, indeed, seen boom and bust but, as Figure 3-3 shows, the amplitude of the economic cycles in terms of farm income has not been much different, perhaps even less dramatic than before.

With regard to depopulation, the Poppers' projections have not been borne out. As shown in Chapter 3 (Tables 3-1 and 3-2) the region has continued to increase in population, although at less than the national rate. Metropolitan areas have grown, while the rural population continues to decline as a fraction of the total, but this is essentially true of the entire country, except where, for quality of life reasons, urbanites move into adjacent rural counties. In actual numbers, the rural population has increased slightly since a low in 1970 (Figure 3.1). Interestingly, urban growth is not confined to only the major metropolitan areas. Smaller and midsize cities that continue to serve the agricultural hinterlands in the Plains have continued to grow (Table 4-5).

The Buffalo Commons concept has many adherents and has stimulated a great deal of discussion in the Plains region and in USA generally. Needless to say, the concept was not received with universal acclaim by residents of the areas concerned. But it has clearly had an impact on individual, corporate and governmental thinking and at this writing a number of grasslands restoration projects, some with large demonstration areas, are underway (e.g., Great Plains Restoration Council).[5]

Licht (1997) finds the Buffalo Commons notion incomplete from his ecological point of view; if its primary aim is the conservation of grassland biodiversity, it will not succeed, he asserts. The region that the Poppers identified for "deprivatization" was almost entirely in the shortgrass or western mixed grass

Table 4-5. Population trend in selected small and midsize Great Plains cities, 1990–2000

City	1990	2000	% Change
Mandan, North Dakota	23,700	25,303	+6.8
Bismarck, North Dakota	83,811	94,719	+13.0
Aberdeen, South Dakota	24,927	24,658	−1.1
Pierre, South Dakota	12,906	13,876	+7.5
Watertown, South Dakota	17,592	20,237	+15.0
Grand Island, Nebraska	39,386	42,940	+9.0
Scottsbluff, Nebraska	35,976	36,951	+2.6
Kearney, Nebraska	25,623	28,211	+10.1
Wichita, Kansas	304,011	344,284	+13.2
Liberal, Kansas	18,743	22,510	+20.1
Dodge City, Kansas	21,129	25,176	+19.2
Enid, Oklahoma	45,366	47,045	+3.7

[5] http://grnc.org/buffalo-commons.html

zones and does not include the tallgrass zones. With or without the Buffalo Commons, other portions of the Plains will continue to produce agricultural surpluses. For Licht, restoring the grasslands does not assure ecosystem restoration. Bison, he states "can be overstocked just as readily as cattle" and fenced-in bison "raise serious health concerns, genetic risks, and ethical questions." Licht argues that, even with movement toward deprivatization and grassland preserves that the Poppers envisioned, agriculture and grass can coexist and livestock grazing will continue on the Plains.

The Buffalo Commons movement rests on the assumption that problems of soil erosion and depletion and degradation of water supplies makes the region's agriculture fundamentally unsustainable. Indeed, soil erosion continues to be of concern in the region although many changes in farm management provide effective means for its control. Developments since World War II and particularly since the 1960s do threaten sustainability of the region's water resources. Groundwater withdrawals exceeding natural recharge rates plus the pollution of surface runoff and groundwater with excess fertilizers (particularly nitrogen) and pesticides are, indeed, cause for concern. A more detailed examination of soil erosion and water supply issues follows. How serious are these problems and what can be done about them?

4. SOIL EROSION ON THE PLAINS—CAUSES AND COUNTERMEASURES

The history of the Great Plains region since European settlement is largely, one of unsustainable land use practices. One may argue that the very opening-up of the region to farming by "breaking of the sod" (replacing the native grass cover with row crops and small grains) was a first strike for unsustainability. Because of its dryness and openness to the sweep of wind, the Great Plains soils are subject to wind erosion. And, because of frequently intense rainfall, water erosion on sloping land is also consequential. As a general rule water erosion is a more serious problem in the eastern portion of the Plains and wind erosion in the west, but both forms of erosion occur throughout the Plains.

Early tillage practices such as cultivating on sloping lands made the soils prone to water erosion. The extension of wheat cultivation westward into the drier zones by itinerant operators termed "suitcase farmers" (Hewes 1973) and others opened the land to wind erosion as did overgrazing the short-grass prairie. The extremity of wind erosion impacts on the Plains region during the "Dust Bowl" years of the 1930s was captured in many iconic photos. Figures 4-2 and 4-3 are good examples.

Water erosion can be controlled by contouring the land so that water runs off along channels of moderate slope, by terracing the slopes and by planting the waterways that conduct water to the bottom of the field with grasses and legumes to hold its soil in place. Erosion of the top layers of soil by rain is a natural process. However, removal of the native grass cover and planting crops (particularly row crops such as corn) in furrows that run up and down the slope greatly accelerates the process.

Wind erosion in the drier portions of the region is also a natural process. Dust storms were encountered by the first European-origin settlers who reached the Plains in the mid-19th century. Bark (1978) cites newspaper accounts of dust storms in Kansas occurring in the 1860s, 1870s, and 1880s. A typical account written in the best of western American dry humor comes from the *Newton KANSAN* of February 24, 1876.

> Last Sunday the Kansas Zephyr was again abroad in the land, and a reasonable quantity of dry and dusty land was aboard the zephyr. It resembled when in good view of the same, across a newly plowed field, or upon a well traveled road, the pictures of a simoon in the desert of Sahara, as depicted in the geographies. The Kansas zephyrs are a promiscuous and pleasant thing, they are. Real estate takes its biggest rise during these times.

Removal of the native grass cover greatly increased the severity of dust storms. Figure 4-4 from a GOES satellite view of the USA on February 23, 1977 shows a dust storm originating near the New Mexico–Texas border.

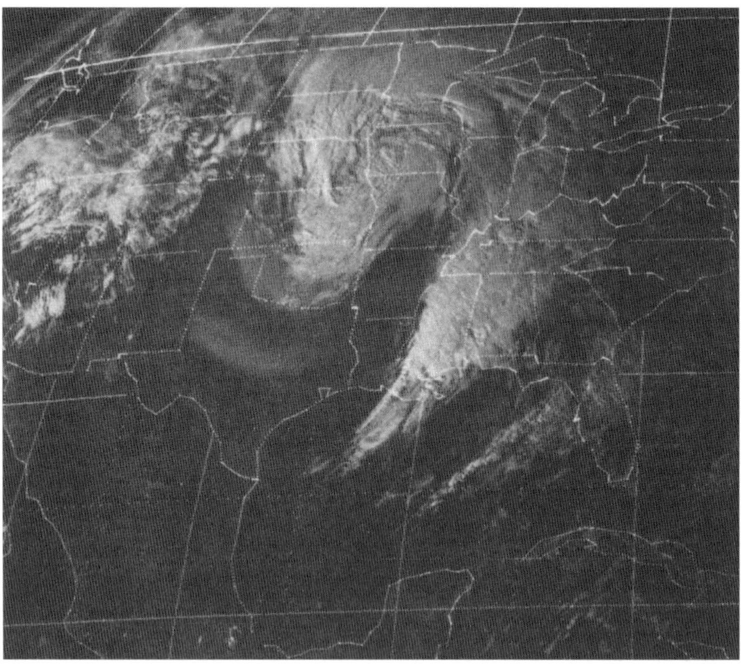

Figure 4-4. Dust storm originating at the New Mexico-Texas border, GOES satellite, February 1977. (Reprinted from Rosenberg 1980)

Rangeland predominated then on the New Mexico side of the border and cropland on the Texas side (Kessler et al. 1978).

Rainfall is insufficient to produce wheat every year In most of the western wheat growing portion of the Plains. Methods of "dry farming" were developed during the late 19th century and early 20th century with the aim of capturing, storing, and conserving every drop of rain falling on the fields in semiarid portions of the Plains. Those methods are well described by Dick (1975).

Among the most popular of the early dryland farming methods was "dust-mulching," pulverizing the upper layer of soil into a loose, finely granular, or powdery layer, and repeatedly tilling it to suppress weed growth (Shannon 1945). The aim was to allow rainfall to penetrate the soil but to interrupt the continuity of the capillaries through which evaporation draws water vapor into the air. Although somewhat effective in conserving moisture, the practice was essentially counterproductive, as a pulverized soil surface is more prone to wind erosion than one that is rough and cloddy.

Because moisture is normally in short supply, fallowing is widely practiced in the region, even today. Fields are planted in the first year and left fallow in the second year to accumulate and store moisture for the next planting. Or, typically, fields are "strip-cropped." Rather than leave large fields fallow, they are divided into strips of equal width. These are usually oriented at approximately right angles to the prevailing wind direction. Half the strips are planted each year and half left fallow to gather and store rainfall for the next season. Each year the planted and fallow strips are rotated. Today, the small grains grown in the northern and western Plains are commonly grown in a "grain-fallow" rotation although, for reasons related to the negative consequences for soil organic matter content, the practice is now being discouraged.

With the passage of time and as the result of research conducted at state, provincial, and Federal Agricultural Experiment Stations, better tillage practices have been introduced to reduce the potential for wind erosion. Strip cropping is only one of these practices. Another is "stubble mulching," defined by the Soil Science Society of America[6] as:

> Stubble mulch: The stubble of crops or crop residues left essentially in place on the land as a surface cover before and during preparation of the seedbed and at least partly during the growing of the succeeding crop.

Stubble mulching is effective in protecting soil from both wind[7] and water erosion.[8]

---

[6] www.soils.org/sssagloss/cgi-bin/gloss

[7] For example, Controlling wind erosion. Kansas Agr. Expt. Station, Ft. Hays Branch Circular 409, May 1977

[8] Mc Carthy, J.R.., Pfost, D.L., and Currence, H.D. Conservation tillage and residue management to reduce erosion. Available at: http://muextension.missouri.edu/explore/agguides/agengin/g01650.htm

Stubble mulching can be considered a precursor of what is now a widespread form of land management—minimum or no-tillage agriculture (sometimes referred to as "conservation tillage"). In this system mechanical plowing and harrowing are eliminated and the crop is seeded directly into thin openings sliced in soil covered with residue of the previous season's crop. Although no-till farming has certain disadvantages (e.g., slower warming in spring; greater requirements for chemical control of weeds), the practice has been shown to be very effective in reducing wind and water erosion in both the drier western wheat lands and wetter corn–soybean land to the east.

The result, over time, of the introduction of the methods described above, particularly of conservation tillage methods has had an effect in reducing the extent of both wind and water erosion in the Plains. Change in the severity of wind erosion from 1982 to 1997 is shown in Figure 4-5. Similarly, change in the severity of water erosion is shown in Figure 4-6.

Another factor to consider is the strong adoption of the Conservation Reserve Program in the Plains states. In 2004 about 14.05 million hectares were enrolled in the National Program. Over 8.08 million of the enrolled hectares or 57.5% of the national total were located in the Great Plains states.[9]

5. WATER PROBLEMS AND COUNTERMEASURES

5.1. Water withdrawals for irrigation

Irrigation is the largest consumptive user of water in the western states and its overuse of the region's water resources poses another threat to sustainability of the agricultural enterprise on the Great Plains. In 1990 Colorado, Montana, and Texas were the largest consumers of irrigation water among the Plains states. Colorado and Montana drew most of their irrigation water from surface sources; in Texas groundwater was the primary source (Table 4-6). Colorado, Montana, and Texas remained among the top four users in 2000 while Nebraska's water withdrawals, mostly from groundwater sources, gave it the dubious distinction of second place. However, as was shown in Chapter 2, groundwater in Nebraska is more abundant and more readily recharged than in Texas. Most of Montana's irrigated land is located west of the Great Plains; significant portions of the irrigated lands in Colorado and Texas are also outside the boundaries of the Plains. As Figure 4-7 clearly shows, Nebraska is now the predominant irrigation state of the region.

Most pertinent to the issue of sustainability are trends in water withdrawals for irrigation. Statistics compiled by the US Geological Survey (e.g., Hutson et al., 2002, on which Table 4–6 is based) show an overall rise in water withdrawals from 1990 to 1995, followed by a larger decrease from 1995 to 2000. Small net increases in withdrawals occurred in Oklahoma and Texas over the decade of the 1990s. However, Nebraska's withdrawals increased by 3.74 km^3 or 44%. All of this increase came

[9] http://www.fsa.usda.gov/dafp/cepd/crp_statistics.htm

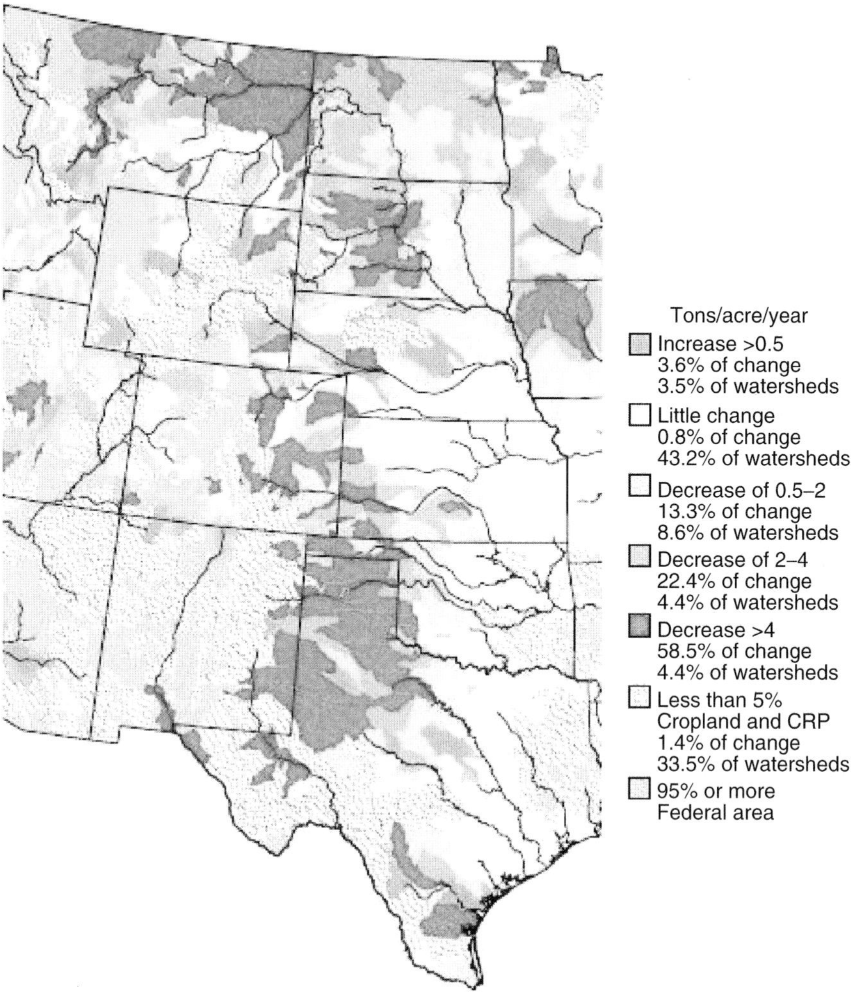

Figure 4-5. Change in average annual soil erosion by wind on cropland and CRP land, 1982–1997. (Source: US Department of Agriculture Natural Resources Conservation Service, http://www.nrcs.usda.gov/technical/land/erosion.html) (*See Color Plates*)

from groundwater, since surface water withdrawals actually decreased in Nebraska during that period.

Figure 4-8 shows change in acreage of irrigated land in the USA between 1997 and 2002. The dramatic increase in irrigated land in the eastern third of Nebraska is evident in this figure as is the equally dramatic decrease in the Texas Panhandle, southwest Kansas, and throughout Colorado.

Overall the Great Plains states used nearly 68 km^3 of water for irrigation in 2000, down 2.1 km^3 from 1990. Of course, individual years vary because of

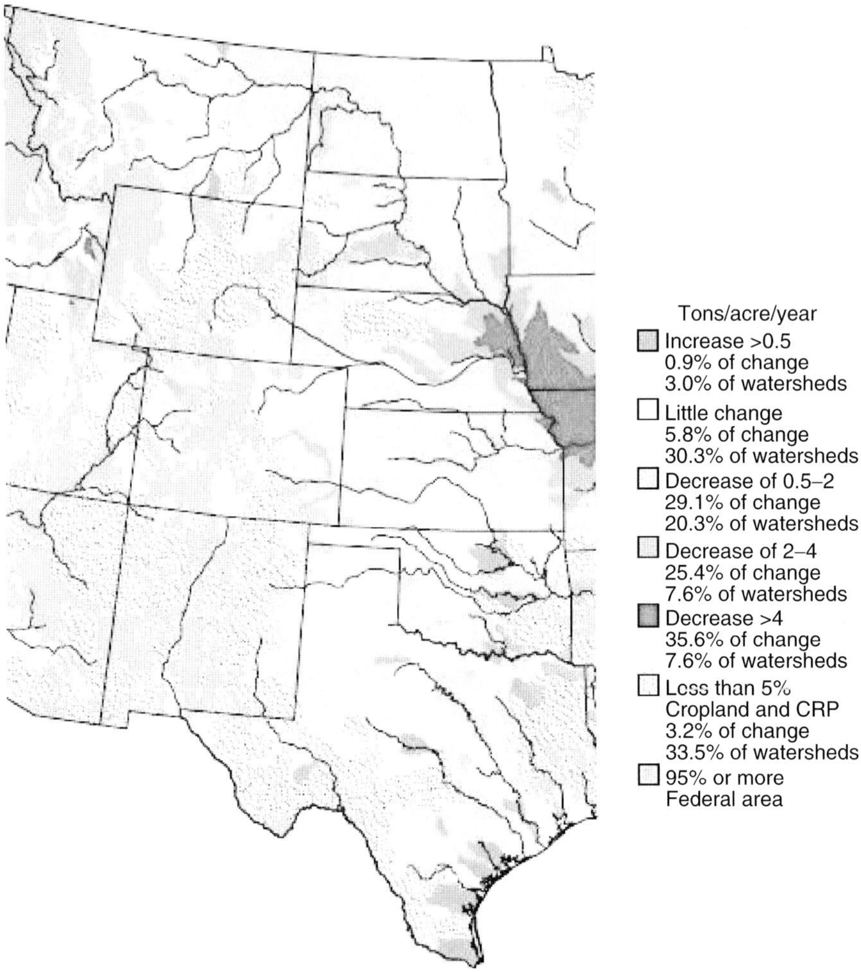

Figure 4-6. Change in average annual soil erosion by water on cropland and CRP land, 1982–1997. (Source: US Department of Agriculture Natural Resources Conservation Service, http://www.nrcs.usda.gov/technical/land/erosion.html) (*See Color Plates*)

weather conditions and in response to farm policy and global market conditions, but the years reported typify observed trends.

5.2. Improvements in irrigation efficiency

Two interrelated factors help to explain current trends in water use for irrigation: rising energy costs of irrigation and improvements in irrigation technology. Good evidence of farmer response to a dramatic rise in energy price is found, for example, in Darmstadter (1993). Following the first oil shock of 1973, farmers—in this

Table 4-6. Irrigation water use by Great Plains States in cubic kilometers per year. (Source: US Geological Survey Circular 1268 and prior publications)

State	1990			1995			2000			Change from 1990 to 2000		
	Ground	Surface	Total	Ground	Surface	Total	Ground	Surface	Total	Ground	Surface	Total
Colorado	3.54	12.46	16.00	2.79	14.80	17.59	2.99	12.83	15.81	−0.56	0.37	−0.19
Kansas	5.51	0.28	5.79	4.37	0.32	4.68	4.74	0.40	5.14	−0.78	0.12	−0.65
Montana	0.12	12.32	12.45	0.11	11.71	11.82	0.11	10.88	10.99	−0.01	−1.44	−1.45
Nebraska	6.02	2.41	8.42	7.99	2.45	10.45	10.26	1.90	12.16	4.24	−0.51	3.74
New Mexico	1.90	2.27	4.17	1.76	2.37	4.13	1.70	2.26	3.96	−0.20	−0.01	−0.21
North Dakota	0.11	0.12	0.23	0.08	0.08	0.16	0.10	0.10	0.20	−0.01	−0.02	−0.03
Oklahoma	0.68	0.15	0.83	1.06	0.14	1.20	0.78	0.21	0.99	0.10	0.06	0.16
South Dakota	0.19	0.35	0.54	0.12	0.25	0.37	0.19	0.33	0.51	−0.01	−0.02	−0.03
Texas	7.73	4.01	11.74	9.03	4.05	13.08	8.99	2.95	11.94	1.26	−1.06	0.20
Wyoming	0.33	9.57	9.90	0.25	8.87	9.12	0.57	5.65	6.22	0.24	−3.92	−3.68
Total	26.15	43.93	70.07	27.56	45.03	72.60	30.44	37.50	67.93	4.29	−6.43	−2.14

Figure 4-7. Irrigated land in the USA, 2002. (Source: 2002 US Census of Agriculture, Map 02-M079) (*See Color Plates*)

case in the four-state Missouri, Iowa, Nebraska, Kansas region—reduced the ratio of energy use to gross product in farming from about 27 to 14 MJ per 1982$. This change was effected in large measure by improvements in irrigation efficiency—defined as the fraction of the applied water that actually recharges the root zone of the irrigated crop—in other words, the fraction not wasted. Better pumps, better maintenance of engines and motors powering the pumps, and improvements in irrigation application technologies explain much of the decreased energy intensity of agriculture.

Furrow and sprinkler irrigation systems predominate in the Plains region. In *furrow irrigation* water is released at the high end of the field into ditches (furrows) lying between the crop rows. Furrows are often 400 m (¼ mile) long or

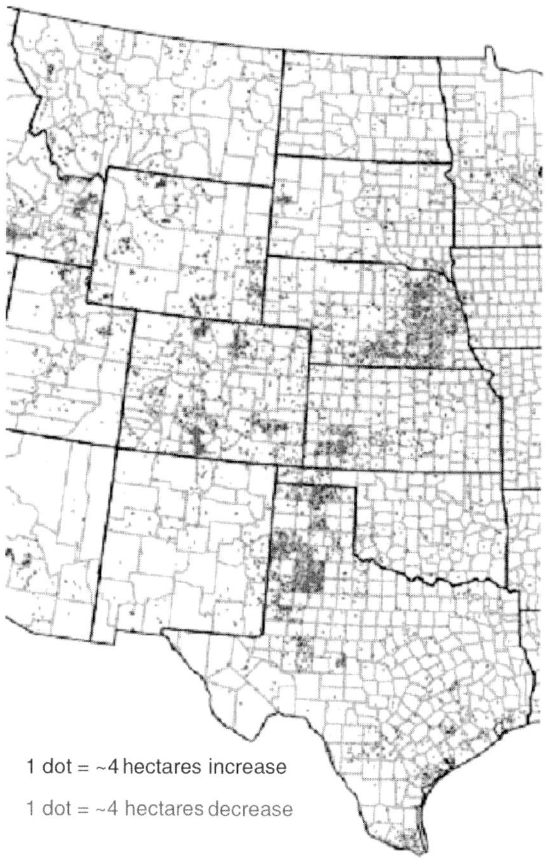

Figure 4-8. Change in area of irrigated land in the USA, 1997–2002. (Source: 2002 US Census of Agriculture, Map 02-M080) (*See Color Plates*)

longer. Depth of water penetration at any particular point in the furrow is a function of (a) the time during which water flows over that point and (b) the rate of water infiltration into the soil. In order to allow for the lower reaches of the furrow to be recharged the general practice has been to allow an excess of water to flow until the end of the furrow has been wetted to the desired depth. Thus, with this practice, a considerable amount of "tail-water" flows wastefully off the field. Furrow irrigation is illustrated in Figure 4-9.

A number of improvements have been made in furrow irrigation practice during the last three decades. First is the use of the "tail-water pit"—a pond dug at the lower end of the field to capture and impound the waste water. From these pits the water is recycled for further use by pumping it back up to the head of the field.

Figure 4-9. Furrow irrigation with gated-pipe. (Source: http://www.wtamu.edu/~crobinson/Irrigation/furgateinfo.html) (*See Color Plates*)

Another important improvement in furrow irrigation has been focused on delivery of water to the field and furrow. Commonly, water is delivered to the field through concrete-lined ditches that are open on top. En route to the field, ditches like these lose a substantial amount of the water to evaporation. Next, water is delivered from the ditch to the individual furrows by means of siphons or weirs. Buried underground pipelines bringing water from the source to the point of application eliminate evaporative losses and increase conveyance efficiency to almost 100%.[10]

Yet another improvement is made when the difficult-to-control siphons or weirs which carry water over the ditch wall into the furrow are replaced with aluminum "gated-pipes" of 15cm or greater diameter supplied directly by the underground delivery system. Figure 4-9 shows gated-pipe lying on the surface at the head of the field and releasing water into individual furrows. Gates on the pipe are opened or closed as needed to regulate flow. Control of flow rate is better with gated-pipe than with either siphons or weirs.

Another important improvement in furrow irrigation is the "surge" system. Rather than run a constant stream of water down the furrow, water is released incrementally. First, enough water is released to wet the upper reaches of the furrow. When that water has been absorbed another dose is released, flowing rapidly over the wetter area to the next dry portion of the furrow. The process is repeated until a measured quantity is sent to service the lowest reach. Water

[10] http://www.wtamu.edu/~crobinson/irrigation/furgateinfo.html

flows are greatly reduced by the use of this technology and much less "tail-water" leaves the field. Most recently, surge systems have been "feedback controlled" with moisture sensors strategically placed in the furrow.

Sprinkler irrigation has long been used on the Plains, but was of relatively limited importance compared with surface methods such as furrow irrigation because of high labor costs involved in moving pipe from place to place in the field. Introduction in the 1960s of the "center-pivot" sprinkler irrigation system revolutionized irrigation on the Plains. Initial capital costs of these systems are high but they are much less labor-intensive in operation than is furrow irrigation.

Center pivot systems consist of a main pipeline 400 or 800 m in length carrying irrigation sprinklers and supported on wheeled towers that move under electrical or hydraulic power (Figure 4-10). The line moves around the field in a circle, pivoting at the center. Water is supplied to the pivot point by pipeline or by a well located at the pivot point.

Most of the US west of the original 13 colonies is laid out in one-mile squares called "sections." A section contains 640 acres; a quarter section contains 160 acres. Thus the pivot pipe may be a quarter of a mile (~400 m) or a half-mile (~800 m) in length and the area irrigated ranges from about 120–130 acres (48–52 ha) in the quarter section to 500 acres (200 ha) in the full section. The area outside the irrigated circle is often used for dryland crop production.

The tower nearest the pivot irrigates the smallest land area and the tower furthest away irrigates the largest area. Therefore, the spacing of the sprinklers and the nozzle size on the mounted sprinklers are adjusted to obtain a uniform water application. One complete rotation of the center pivot can take as long as 2 or 3 days. Given a typical irrigation application of 10 cm: (~4 inches) and assuming 100% irrigation efficiency the smaller system will dispense about 50,000 m^3 of water during a complete cycle and the larger system about 200,000 m^3. Actually these volumes would be greater in high-pressure systems since evaporative losses and wind drift reduce irrigation efficiencies to more like 80–90%.

Before the first "oil shock" in 1973, energy costs for irrigation were low. But with the steep increases in energy prices after that time irrigation engineers focused on reducing costs by improving irrigation efficiency. One approach taken was to reduce the pressures at which the sprinklers operate in order to reduce wind drift and evaporative water losses. In another adjustment made to reduce evaporative losses the sprinkler heads are turned to face downward rather than upward.

Drip-irrigation: One might ask why the highly efficient "drip irrigation" system is not widely used on the Plains in view of the importance of water conservation in the region. Drip irrigation systems consist of small plastic tubes with closely spaced orifices that are laid alongside or even buried in the soil adjacent to the plants to be watered. Water is provided at a rate intended to recharge the soil as it is being extracted by evapotranspiration. Drip systems are probably the most "irrigation efficient" of all irrigation systems and their cost and upkeep is justified when used to water high-value crops such as orchards, vineyards,

Figure 4-10. Center pivot irrigation system. (Source: http://www.ars.usda.gov/is/graphics/photos/oct00/k9072-1.htm) (*See Color Plates*)

vegetables, nursery stock, and so on. However, the drip system is not well suited to the large fields of relatively low per-acre-value crops currently grown in the Plains region.

The use of "sensor-guided" computer-controlled irrigation scheduling technologies has also contributed importantly to improvement in energy and irrigation efficiencies. In this approach sensors that monitor soil moisture in the field and automated weather stations that monitor net radiation, air and soil temperature, atmospheric humidity, windspeed, and precipitation provide a constant flow of data for computation of evapotranspiration rate. Combining these calculations with weather forecasts, computer programs project when and how much water will be needed to maintain best growing conditions in the field.

Computer-guided irrigation scheduling began in the 1960s and has increased in sophistication ever since as sensors and within-field electronic communications have improved and as desktop and laptop computers have become as ubiquitous as the pitchfork on modern, well-managed farms in the USA, Canada and elsewhere in the developed world.

Each of the technologies described above has had an impact in conserving irrigation water and optimizing the efficiency of its application. Further improvements are in the offing: surveillance by satellites or unmanned aircraft of crop conditions including moisture stress, GPS-controlled changes in sprinkler rates consistent with within-field differences in soil characteristics, and so on. It seems reasonable to assume that worsening water shortages and continually rising energy prices will prompt yet more imaginative technological improvements for the irrigation enterprise on the Plains and throughout the world.

6. SUSTAINABILITY ON BALANCE

The earlier portions of this chapter describe the nature, extent, and productivity of contemporary Great Plains agriculture and current arguments concerning its sustainability. At this point no definitive answer can be given as to whether that agriculture is becoming more or less sustainable. From the point of view of land management and water conservation practices developed and implemented over the last two or three decades, the indication is *probably more sustainable*. An expert on sustainable agriculture, Professor Charles Francis of the University of Nebraska–Lincoln[11] attributes improved agricultural efficiency to improved use of resources through conservation—and no-till farming, better use of available water, good plant breeding programs that have brought a high level of drought tolerance into corn, greater use of crop rotations in the farming systems of the region, and improved irrigation efficiency. Professor Francis foresees an end to corn irrigation on the Plains, except for seed production in the next 25 years, as the result of growing demands and competition for water. The "precious irrigated acres" will be used for crops much higher in value than corn. Whether the increasing demand for ethanol made from corn or other feedstocks will negate Francis' prediction remains to be seen.

Another important point needs be made here is resilience, and hence sustainability, of the current agricultural enterprise, with all its new varieties and technologies, has yet to be tested under long-duration climatic extremes such as those of the 1930s droughts. Probability tells us that such extremes must someday recur. Nor can it be ruled out that even more extreme climatic conditions may occur because of exacerbating effects of global warming, which is the subject of the chapter to follow.

---

[11] Personal communication, February 22, 2006.

REFERENCES

Bark, L.D., 1978, History of American droughts, in Rosenberg, N.J. (ed.), *North American Droughts*, American Association for the Advancement of Science, Selected Symposium Series, Westview Press, Boulder, Colorado, pp. 9–23.

Bruntland, G.H. (ed.), 1987, *Our Common Future: The World Commission on Environment and Development*, Oxford University Press, Oxford, 398 pp.

Darmstadter, J., 1993, Climatic change impacts on the energy sector and possible adjustments in the MINK region. *Clim. Chang.* **24**:117–129.

Dick, E., 1975, *Conquering the Great American Desert: Nebraska*, Vol. XXVII, Nebraska State Historical Society Publications, Nebraska, 456 pp.

Francis, C.A., 2004, Sustainable agriculture, in Wishart, D.J. (ed.), *Encyclopedia of the Great Plains*, University of Nebraska Press, Lincoln, Nebraska, pp. 53–54.

Hewes, L., 1973, *The Suitcase Farming Frontier: A Study in the Historical Geography of the Central Great Plains*, University of Nebraska Press, Lincoln, Nebraska, 281 pp.

Hutson, S.S., Barber, N.L., Kenny, J.F., Linsey, K.S., Lumia, D.S., and Maupin, M.A., 2002, *Estimated Use of Water in the United States in 2000*, US Department of Interior Geological Survey Circular 1268, 52 pp.

Kessler, E., Alexander, D.Y., and Rarick, J.F., 1978, Duststorms from the US High Plains in late winter 1977—search for cause and implications, *Proc. Oklahoma Acad. Sci.* **58**:116–128.

Licht, D.S., 1997, *Ecology and Economics of the Great Plains*, University of Nebraska Press, Lincoln, Nebraska, 225 pp.

Popper, D.E. and Popper, F.J., 1987, The Great Plains: from dust to dust, *Planning* **53**:12–18.

Popper, D.E. and Popper, F.J., 1999, The Buffalo Commons: metaphor as method, *Geogr. Rev* **89**:491–510.

Rosenberg, N.J. (ed.), 1980, *Drought in the Great Plains: Research on Impacts and Strategies*, Water Resources Publications, Littleton, Colorado, Frontispiece.

Shannon, F.A., 1945, *The Farmer's Last Frontier: Agriculture, 1860–1897*, Vol. 5: *The Economic History of the United States*, Harper & Row, New York, 434 pp.

CHAPTER 5

THE WILDCARD OF CLIMATE CHANGE

1. INTRODUCTION

To this point we have examined the natural resources of the NAGP, its economy and the nature of current land use in the region. As explained in foregoing sections, land use on the Plains has evolved as the result of historical forces in response to changing demographics, economic conditions, and public policy. But until now public policy, rules for managing the Missouri River dams, drought mitigation programs, etc., as well as general and public expectations, continue to rest on the assumption that the climate of the region, while encompassing large and sometimes dramatic day-to-day and year-to-year variability, is essentially stable and fundamentally unchanging.

Yet all of us have memories that lead to such thoughts as ..."winters sure ain't what they used to be"...or...."spring doesn't last as long as it used to"...or..."the rains are getting heavier and heavier"..., and so on. Such musings imply that the climate is changing. However, until the last decade or so there was little evidence that what some perceive as "change" actually lies outside the range of normal climate variability. But an emerging body of evidence that squares well with this theory indicates that climate is, indeed, now changing—globally, not only in the Plains. And at least one driver of climate change, perhaps the preeminent cause, is "anthropogenic", i.e., the result of human activity.

We know that the atmospheric concentrations of carbon dioxide (CO_2), methane (CH_4), nitrous oxide (N_2O) and other radiatively active trace gases are rising due to the combustion of fossil fuels, cement manufacture, and changes in land use and management, and that these gases strengthen the Earth's natural "greenhouse effect". This process can only lead to a warming of the lower atmosphere, the land and its water surfaces and, more slowly, the oceans. Global warming is, itself, a climate change but other aspects of climate—precipitation, winds, and currents must also change as the Earth warms.

What is the extent of the warming so far? For more than a century, maximum and minimum air temperatures have been measured daily at about 1.8–2.0 m above ground surface at many thousands of sites around the world. Records compiled from these stations and from lake and sea observations mostly in the Northern Hemisphere show that the globe has, in fact, warmed to the extent of

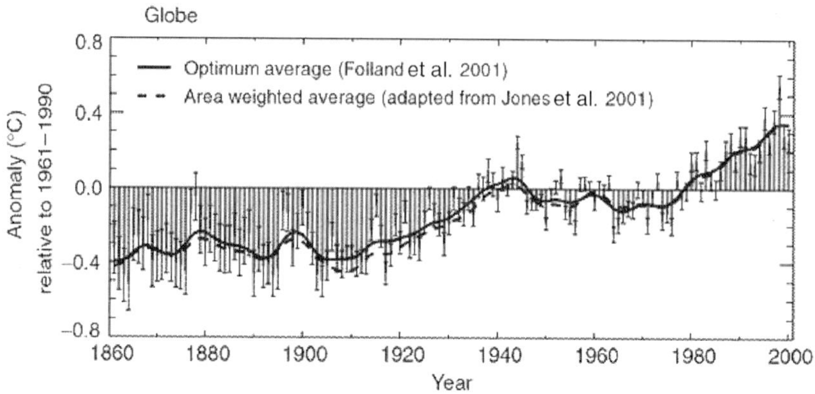

Figure 5-1. Combined annual land-surface air and sea-surface temperature anomalies (°C) 1861–2000, relative to 1961–1990. Two standard error uncertainties are shown as bars on the annual number (IPCC Third Assessment Report 2001)

0.6°C during the last 140 years. The warming trend is shown in Figure 5-1 taken from a report of the Intergovernmental Panel on Climate Change (IPCC) (2001a). This figure shows a long upward trend in temperature interrupted by a decrease between 1940 and 1975. Thereafter, a warming trend has dominated the record. Climatologists at NASA's Goddard Institute for Space Studies (GISS) have noted that the highest global annual average surface temperature in more than a century was recorded in their analysis for the 2005 calendar year. The records show that the warmest 5 years since the 1890 have been, in order of descending temperature, 2005, 1998, 2002, 2003, and 2004.[1] It has also been observed (e.g., Karl et al. 1993, 1996) that daily minimum temperatures have risen more than maximum temperatures. Nights are warming more than days, which is consistent with the theory of "greenhouse warming" described below.

Indicators of change documented by the IPCC (2001b) include shrinkage of the area of sea ice in the Arctic, increasing depth of the active layer of soil and organic materials overlying permafrost in Alaska, and lengthened agricultural growing seasons in North America and Europe in the latter half of the 20th century. Two seminal papers on the ecological effects of warming appeared in *Nature* in 2003 (Root et al. 2003; Parmesan and Yohe 2003). Root et al. selected 143 studies from thousands examined that met certain criteria and found a consistent temperature-related shift "....in species from mollusks to mammals." Parmesan and Yohe documented significant range shift averaging 6.1 km

[1] http://www.nasa.gov/vision/earth/environment/2005_warmest.html

poleward per decade and a significant mean advancement of spring events of 2.3 days per decade in 279 species.

The temperature decrease between 1940 and 1975 seen in Figure 5-1 provides an easy way to emphasize the notion that climate is always changing and has always done so because of natural phenomena uncontrollable by man, but a growing body of evidence points to significant anthropogenic changes now occurring in climate that are likely to become increasingly evident during the course of this century. Among these are a warming of the lower layers of the atmosphere, particularly in the higher latitudes; an intensification of the hydrologic cycle leading to more evaporation and precipitation but with a geographical distribution different from today's; a year-round decrease in the average extent of the arctic ice cap and its possible disappearance in summer; and a rise in sea level that may be great enough to force abandonment of many low-lying areas or necessitate the construction of expensive protective systems. Why might all this happen?

The following sections of this chapter briefly examine the mechanisms of climate change, projections of how, given varying severities of global warming, the climate of the NAGP could change over the course of this century and what the impacts of such changes might be for important sectors of the region's economy.

2. THE GREENHOUSE EFFECT AND GLOBAL WARMING

2.1. The energy balance

Figure 5-2 from Kiehl and Trenberth (1997) is a schematic description of the Earth's annual and global mean energy balance. Mean flux density of incoming solar radiation at the top of the atmosphere is 342 W m^{-2}. Of the incoming solar radiation, which is primarily in the visible waveband, the atmosphere absorbs 67 W m^{-2}. Clouds, aerosols and atmospheric gases reflect 77 W m^{-2} to space and the Earth's surfaces reflect another 30 W m^{-2}. The surface absorbs 168 W m^{-2} of the incoming solar radiation. The Earth-atmosphere system must dispose of the incoming energy or the planet would warm uncontrollably. The surface disposes of energy by a number of mechanisms: warming the air coming in contact with it (24 W m^{-2}) and by evapotranspiration (ET)—direct evaporation of water from the soil and free-water surfaces and by transpiration—evaporation at the leaf-surface of water drawn through plants (78 W m^{-2}). The surface also emits (on average) 390 W m^{-2} by thermal infrared radiation of which 40 W m^{-2} passes through what is called the "atmospheric window"—a portion of the spectrum from about 8 to 12 microns in which water vapor is not a strong absorber of longwave radiation. Back radiation to the surface from the atmosphere in the thermal waveband is 324 W m^{-2}. The outcome of the exchanges of thermal radiation is an outgoing flux density of longwave radiation at the top of the atmosphere of 235 W m^{-2} which, together with the reflected radiation of 107 W m^{-2}, balances the incoming solar radiation.

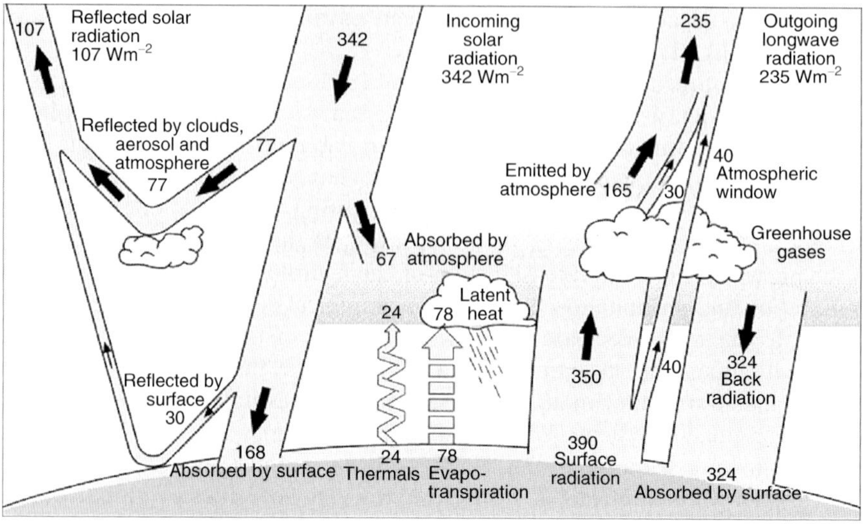

Figure 5-2. The Earth's annual and global mean energy balance (Kiehl and Trenberth 1997; Reprinted from IPCC Third Assessment Report 2001a)

2.2. The greenhouse gases

Our particular concern here is with the radiation passing through the atmospheric window. Natural constituents of the atmosphere—water vapor, CO_2, CH_4, N_2O, and ozone (O_3) and some manufactured substances such as the chlorofluorocarbons (CFCs) are partially opaque to the longer wavelength thermal radiation and trap a portion of it. These gases, except for water vapor, have strong absorption peaks in the "atmospheric window."

A portion of the energy absorbed by these "greenhouse gases" is retained in the lower layers of the atmosphere, raising its temperature. Since the late 19th century, scientists have speculated that because of their strong absorption of infrared radiation, especially in the atmospheric window, the rising concentrations of CO_2 and the other radiatively active trace gases cause a warming of the lower layers of the atmosphere. Warming caused by these gases also increases the atmosphere's capacity to retain water vapor, a positive feedback toward warming. This phenomenon has been likened to the process that occurs in greenhouses: the glass permits solar radiation to penetrate, but it absorbs infrared radiation emitted by the soil and plants within. Although the analogy is defective, the process of warming in the lower layers of the atmosphere caused by the infrared absorptive behavior of CO_2 and the other radiatively active trace gases mentioned above has come to be known as the "greenhouse effect."

Carbon dioxide: It is not known for certain what the concentration of CO_2 was before the industrial revolution, although the record of sporadic air analyses from

1870 on and samples of air trapped in glacial ice suggest a value of ~280 parts per million by volume (ppmv). From that time on, as coal became increasingly important as the source of energy to power the industrializing society, CO_2 concentration began to increase. As a graduate student in meteorology and soil science in the late 1950s, I learned that the concentration had already risen to 315 ppmv. As this book is written (2006) atmospheric CO_2 concentration is approaching 380 ppm (Blasing and Jones 2005). Thus, the atmospheric loading of this important greenhouse gas has increased by more than a third since the beginning of the industrial revolution. The emissions of CO_2 in 2002 from fossil fuel combustion, cement manufacture, and land use change injected ~6.975 Gt carbon[2] into the atmosphere, up 2% from the previous year (Marland et al. 2005).

Methane: Methane (CH_4) is the greenhouse gas next in importance to CO_2 in radiative forcing. Although emitted in much smaller quantities than CO_2 (millions rather than billions of metric tonnes per annum),[3] the CH_4 molecule has a greenhouse warming potential 23 times greater than that of CO_2. CH_4 concentration had more than doubled from a preindustrial (before 1750) concentration of ~0.7–1.73 ppmv and 1.85 ppmv in 2003. CH_4 is a fossil fuel and can leak from oil wells and natural gas processing and transport facilities. It is also emitted under anaerobic conditions from swamps, rice paddies, landfills, and by ruminant animals and termites. Emissions were in the range of 600 Tg of CH_4 in the late 1990s (IPCC 2001a).

Nitrous oxide: Nitrous oxide (N_2O) is released from soil to the atmosphere during the nitrification process that occurs in soil under aerobic conditions. The use of nitrogenous fertilizers increased greatly after World War II, and is primarily responsible for the increase from its preindustrial atmospheric concentration of 270–318 parts per billion (ppb) in about 2003. However, N_2O is also emitted as the result of denitrification—a process that occurs in waterlogged soils. It has been suggested that the drainage of such soils and alterations in their acidity may have reduced this source of natural nitrogenous emissions to some degree. N_2O is also a more effective greenhouse gas than CO_2. In 1990 N_2O was increasing at the rate of 0.25% per annum. Its warming potential per molecule is ~296 times that of CO_2. Emissions in the late 1990s were in the order of 16.4 Tg (IPCC 2001a).

Clorofluoromethanes: The clorofluoromethanes or "Freons" (CFCs 11 and 12), used as refrigerants, propellants, and for cleaning electronic components are human-made, having no natural sources. Although better known for their role in erosion of stratospheric ozone layer to which they deliver chlorine atoms that catalyze the photolytic destruction of O_3, the Freons are also extremely strong infrared absorbers. In the late 1990s concentrations of CFCs 11 and 12 were 253 – 256 and 542 – 546 parts per trillion (ppt), respectively (Blasing and Jones 2005). Molecules of these gases have greenhouse warming potentials 4,600 and

[2] Billions of metric tonnes = Gt = gigatonne = petagram = 10^{15} g.

[3] Millions of metric tonnes = Tg = terragram = 10^{12} g.

10,600 greater than that of CO_2. Production of CFC 11 and 12 was banned by the Montreal Protocol,[4] and its concentration is no longer rising; indeed it has begun to fall. However, some of the chemicals that are intended to replace the Freons, while nonthreatening to the ozone layer, are even stronger as greenhouse gases.

2.3. The CO_2-fertilization effect

Although CO_2 is the greenhouse gas of the greatest immediate concern with regard to global warming because of the immense quantities being emitted, the increase in its atmospheric concentration can have a potentially positive effect as well. CO_2 is the substance from which plants synthesize the basic sugars, building blocks of all plant products, through the process of photosynthesis. The increase in its atmospheric concentration affects plants in two ways. Photosynthesis is increased in the C-3 plants, those having a 3-carbon intermediate molecule in the photosynthetic pathway. The C-3 plants include the legumes, small grains, cool-season grasses, most root crops and trees. Photosynthesis is only slightly affected in the C-4 plants, tropical grasses such as corn, sorghum, millet, sugarcane, and some warm-season grasses, which have a more efficient photosynthetic mechanism than the C-3 group. But elevated CO_2 concentration has the effect of partially closing the stomates (pores) of plant leaves and stems in both C-3 and C-4 plants, making the diffusion of water vapor into the air more difficult. This results in decreased transpiration and conservation of soil moisture. Both groups of plants experience an improvement in their water-use efficiency (WUE, the ratio of photosynthetic production and water consumption). The degree to which this "CO_2-fertilization effect" might offset stresses on crops caused by climatic change or might augment beneficial effects of climate change, should such occur, is evaluated in subsequent portions of this chapter.

3. CLIMATE CHANGE SCENARIOS

Understanding of how climate change might affect natural processes such as plant growth, ET, and ecological functioning in general, requires models that can mimic the processes involved. A few such physical and biological process models are described below. But before these models can be employed, information must first be provided on how the climate might actually change in the future. A number of techniques are used to generate "scenarios" of climate change. Among these are: climatic analogues, statistical regressions, and general circulation models.

---

[4] Montreal Protocol (Protocol on substances that deplete the ozone layer) is a treaty signed by 25 nations in 1987. The protocol set limits on the production of the CFCs, halons, and other substances that release chlorine or bromine into the upper atmosphere where ozone is concentrated. The protocol has been amended several times and 168 nations are now signatories (source: http://www.factmonster.com/ce6/sci/A0833884.html).

3.1. Climate analogues

Climatic phenomena of the past provide a basis for formulating scenarios of future climate change. The actual climate record of the drought era of the 1930s in a portion of the Great Plains and adjacent states (Missouri, Iowa, Nebraska, and Kansas) was used, in a study reported by Rosenberg (1993), to create scenarios of climate change for the region to the year 2030. These scenarios, essentially a replay of the drought of the 1930s (the "dirty-thirties") climate, were used to evaluate potential impacts on agriculture, forestry, water resources, energy supply, and demand in the region as it was in 1990 and as it might be in 2030. The climate analogue records were applied to process models and other tools in order to accomplish this.

Glantz and Ausubel (1984) posed another sort of climate analogue for the Great Plains, using known and anticipated impacts of the depletion of the Ogallala aquifer (Chapter 2) as a guide to what might happen in the region if, as anticipated, climate change depletes water resources in the region.

A general weakness of the analogue approach is, of course, that the climate events or deviations from normal in the past may be quite different from what might occur under greenhouse-forced climate change. For instance, the hot, dry conditions of the Dust Bowl days may not be repeated under climate change; a hot, wet future for the Dust Bowl region is not out of the question.

3.2. Statistical regression

The climate record has also been used to establish sensitivity of various crops to temperature and precipitation fluctuations by means of statistical regression. Agronomists (e.g., Thompson 1986) have established how, over a long period of time, the final yield of corn and soybeans in Iowa is affected, say, by a 1 – 3°C warmer than usual May, June, July mean temperature, or by a 10 – 20% wetter or drier conditions in these months. The relationships established are then used in regression equations to project the effects of such changes should they become the long-term means in the future. Newman (1980) and Blasing and Solomon (1983) were among the first to use the regression approach. While the regression approach provided useful insights about climate change impacts on crop production, projecting statistical relationships developed under a limited range of current conditions far out of that range is an uncertain practice.

3.3. General circulation models

The general circulation model (GCM) is a global, three-dimensional computer model of the climate system which can be used to simulate human-induced climate change. It has become the primary tool used by climatologists to analyze the effects of such factors as reflective and absorptive properties of atmospheric water vapor, greenhouse gas concentrations, clouds, annual and daily solar heating, ocean temperatures, and ice boundaries.[5]

Essentially, the GCM divides the surface of the globe into a two-dimensional grid, typically several hundreds of kilometers on a side. The atmosphere above each two-dimensional grid box is divided into 10–20 layers reaching to about 35 km, effectively the top of the atmosphere. In coupled ocean-atmosphere models, the oceans are similarly subdivided into grids and layers. The equations of motion, the radiation balance and the properties of the atmosphere determine the dynamics of the atmosphere or ocean within each three-dimensional cell. Each cell exchanges momentum, heat, and water vapor with its neighbors. It is no wonder, then, that the computational requirements of GCMs, with their tens of thousands of cells and the complex physical phenomena they simulate, have been a major factor leading to development of super-computers.

The GCM can be applied in the short-term (years) to explore, for example, the effects of volcanic eruptions; the relatively long-term (decades to centuries) to explore the potential effects of the increasing atmospheric burden of aerosols, greenhouse gases, and other substances; or the truly long-term (millennia) to explore the effects of changes in solar luminosity and/or variations in the Earth's orbit around the sun. There are perhaps two dozen GCMs presently being used to project the timing and geographic distribution of greenhouse-forced climate change.

Although simpler models have been used to provide globally or regionally averaged estimates of the climate response to greenhouse warming, the scientific consensus is that only GCMs, possibly in conjunction with regional models nested within them (see below), have the potential to provide the geographically and physically consistent estimates of regional climate change that are required in impact analysis.[6]

The scale of the grid box in a GCM is usually too large to provide information that can be directly applied to a farm or even a county or watershed. In regions of complex topography—say Washington State—a single grid box may encompass range, or wheat land, mountains, and ocean, so that the average temperature or precipitation change projected has little value for impacts assessments. Some researchers (e.g., Georgi and Mearns 1991; Georgi et al. 1998; Brown et al. 2000) use smaller, more geographically detailed models "nested" within the larger grid cell and driven by its projected climate changes to provide information at a more usable scale.

What do the GCMs predict for the Plains region? There is no definitive answer to this question since, as mentioned above, there are many GCMs in current use. Because of differences in the ways that physical processes are parameterized and because of differences in computational strategies employed, agreement among the

---

[5] eobglossary.gsfc.nasa.gov/Library/glossary.php3

[6] IPCC Data Centre, ipcc-ddc.cru.uea.ac.uk/ddc_gcm_guide.html

GCMs is not always gratifying. For example, among six of the most widely used GCMs (BMRCD2, CCC1D2, ECH4D2, GFDLD2, GISSD2, and HAD3D2)[7] correlation coefficients with respect to their projections for normalized annual temperature change range from a low of 39 to a high of 83 where 100 would represent perfect agreement. The situation with respect to normalized annual precipitation change is considerably worse with a low value of 3 to a high of 39.[8]

Nonetheless, it is possible to see in Figures 5.3a and b the broad outlines of the climate futures projected for the Great Plains by a number of the better-known and accepted GCMs. These figures have been assembled by means of a computer model SCENGEN (scenario generator) developed by the Climate Research Unit of the University of East Anglia (Hulme et al. 1995) in cooperation with National Center for Atmospheric Research in Boulder, Colorado.[9]

3.3.1. The US National Assessment

GCMs have been used in some recent assessments of climate change impacts on the USA. A major assessment organized by agencies of the US government reported its findings in the "National Assessment" (USGCRP 2002). This report was organized to deal with "mega-regions" of which the Great Plains was one. Two GCMs—the Canadian Global Coupled Model (CGCM, Flato et al. 2000; Canadian model hereafter) and the Hadley Centre model (HadCM3, Johns et al. 1997; Hadley model hereafter)—were used to project climate change in the 21st century. The models were also used to retrospectively project the climate changes of the 20th century.

The Plains region as a whole warmed by 0.5 to 1.0°C during the 20th century. The Canadian model shows warming in the order of 5.5°C and the Hadley model shows warming of about 2.3°C in the 21st century. During the 20th century, precipitation decreased by 10% in eastern Montana, North Dakota, eastern Wyoming, and Colorado, and increased by more than 10% in the eastern portion of the Great Plains. Also the snow season ended earlier in spring because of greater seasonal warming in winter and spring. The Canadian model projects decreases in the 21st century of as much as 30 – 40% in precipitation in the southern Plains, and increases of 20% in the northern Plains. The Hadley model, on the other hand, projects increases in the range of 20% in almost all of the US Plains with some modest decreases east of the Rockies.

3.3.2. GCMs in the JGCRI study

Another assessment aiming to derive impacts of projected climate changes on agriculture, water resources and irrigation in the conterminous USA was recently

---

[7] Bureau of Meteorology Research Centre (Australia); Canadian Centre for Climate Modeling and Analysis; Max Planck Institute (Germany); Geophysical Fluid Dynamics Institute (USA); Goddard Institute for Space Studies (USA); and Hadley Centre (Gt. Britain).

[8] SCENGEN website http://www-pcmdi.llnl.gov/projects/cmip/cmip_abstracts/wigley03.pd

[9] Latest updated versions at http://www.cru.uea.ac.uk/~mikeh/software/scengen.htm

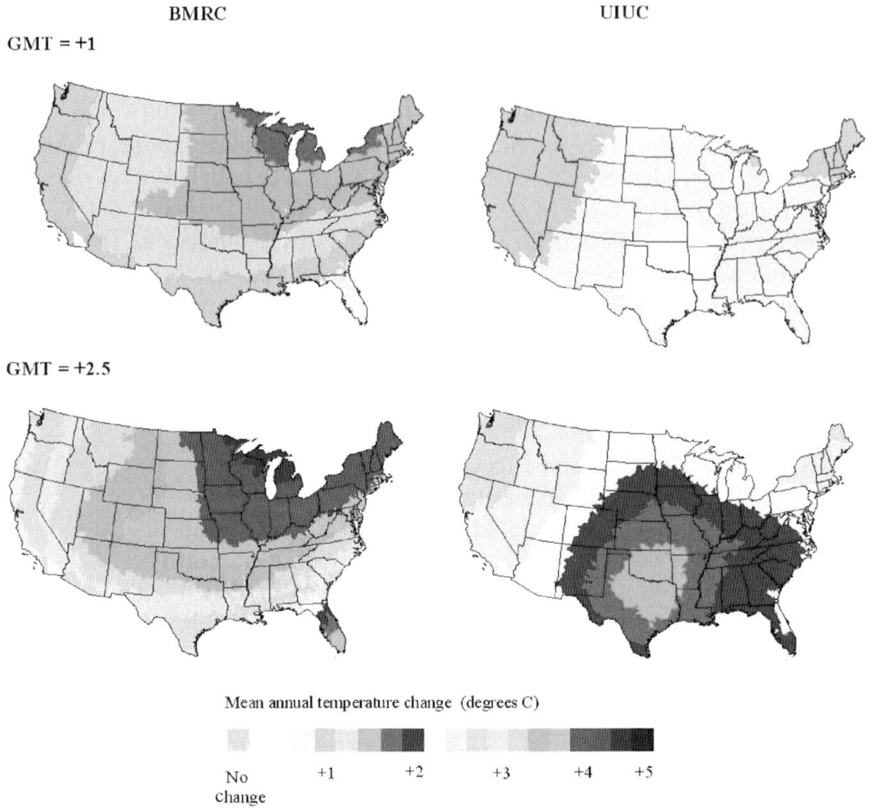

Figure 5-3a. Mean annual temperature change from baseline for the BMRC and UIUC GCMs used in the JGCRI study (Source: Smith et al. 2005) (*See Color Plates*)

reported in Rosenberg and Edmonds (2005). This study was conducted by scientists at the Joint Global Change Research Institute (JGCRI)[10] and made use of two other GCMs—the Australian Bureau of Meteorology Research Centre (BMRC) model (McAveney et al. 1991), and the University of Illinois, Urbana–Champagne (UIUC) model (Schlesinger 1997). These models illustrate the range of differences in climate change projections that can be found among GCMs (Figures 5-3a and b).

Virtually all GCMs indicate some degree of warming, but the range of change in monthly, seasonal, and annual temperatures can be considerable. The UIUC and BMRC models used by Thomson et al. (2005a) in the JGCRI study project

---

[10] A collaboration of the Pacific Northwest National Laboratory and the University of Maryland–College Park, USA.

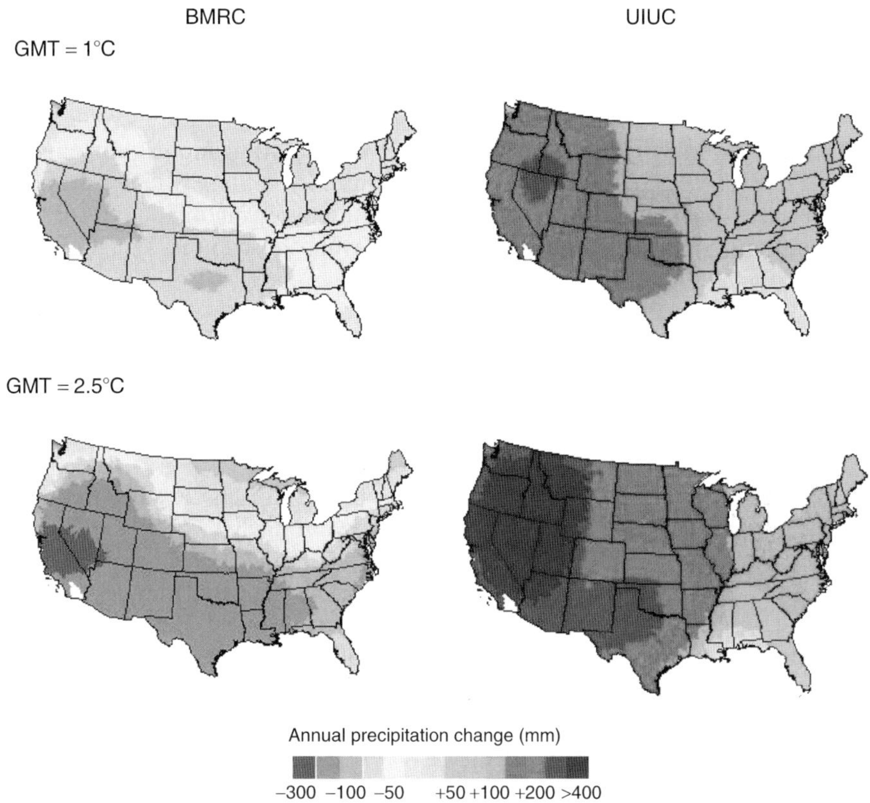

Figure 5-3b. Mean annual precipitation change from baseline for the BMRC and UIUC GCMs used in the JGCRI study (Source: Smith et al. 2005) (*See Color Plates*)

annual mean temperature change in eastern Nebraska to range from <1°C to ~2°C, respectively, when global mean temperature has risen by 1°C. With global mean warming of 2.5°C, these same models project mean annual temperature at the same location to range from slightly under 2°C to 4°C.

Differences in precipitation projected by the GCMs can be considerably greater, and ultimately of greater importance. In the Thomson et al. (2005a) study the UIUC and BMRC models project, respectively, a 50% increase and a 50% decrease in precipitation in eastern Nebraska with global mean temperature change of 1°C. With GMT of 2.5°C UIUC projects a 100% increase; BMRC continues to project a 50% decrease.

Seasonal differences can be even more significant. BMRC projects temperatures higher than UIUC by 2.09°C, 1.60°C, 1.05°C, and 1.35°C for winter, spring, summer, and fall in the Missouri River Basin at GMT = 2.5°C. In the Arkansas Basin these differences are 2.02°C, 2.58°C, 1.06°C, and 1.04°C for the same seasons. Differences in seasonal precipitation projected by UIUC and BMRC are 28%, 33%, 164%, and 113% for winter, spring, summer, and fall, respectively in the Missouri River basin and 91%, 22%, 184%, and 138% in the Arkansas.

The uncertainties inherent in using general circulation models for climate change impact study should evoke a sense of humility in those who use them for that purpose. Keeping those uncertainties and limitations in mind, and for lack of a better alternative, we proceed to examine what those impacts might be for the NAGP.

4. IMPACTS OF PROJECTED CLIMATE CHANGES ON PLAINS AGRICULTURE AND WATER RESOURCES

4.1. Introduction

Many researchers who have dealt with climate change impacts conclude that agriculture is one—perhaps the major—economic sector likely to be affected in the USA by climatic change (e.g., Rosenberg 1982; National Academy of Sciences 1992; IPCC 2001b; Reilly et al. 2003). There are many possible modes of agricultural adaptation to climatic change if it cannot be avoided (Rosenberg 1992). These include introduction of new, better-adapted crops, development of new cultivars for current crops, changes in tillage practices to optimize management in response to changes in season length and other fairly obvious adjustments. Irrigation would be the most effective way to compensate for rising temperatures, greater ET and, in some regions, reduced precipitation. This assumes, of course, availablity of water to irrigate where dryland yields fall below profitable levels. But will the water actually be there?

Studies based on the use of process models have been conducted for agricultural systems throughout the world and for most of its crops. Summaries of such studies are to be found in the first three IPCC Assessment Reports (1991, 1996, 2001)[11] and in many national assessments—the most relevant of which for this book are the US National Assessment (USGCRP 2001) and the JGCRI study (Rosenberg and Edmonds 2005) cited above, each of which used a different pair of GCMs to drive process models of crop growth and yield and regional water supply.

4.2. Results of the National Assessment

Assessments of climate change impacts are developed using process models driven by GCM projections. In the National Assessment which involved analysts from a number of research centers, a range of process models were used for

[11] Accessible through http://www.ipcc.ch/pub/reports.htm

this purpose. The general outcome of their model runs with respect to the Great Plains is as follows: applying the climate scenarios stemming from the Canadian and Hadley GCMs to process models for major crops and considering crop and livestock production weighted by prices, the National Assessment found that regional production changes in 2030 and 2090 relative to current production differed greatly by region and by GCM. Under the Canadian model the northern Plains showed a small loss in 2030 where no adaptation was attempted and a small gain with adaptation. Loss was in the order of 20% with no adaptation in 2090 but adaptation increased production by 20%. For the southern Plains under the Canadian Climate model, regional production suffered losses in 2030 or 2090, only partially alleviated by adaptation.

Under the Hadley model of climate change both the northern and southern Plains show increased regional production. In the northern Plains the increases are roughly 10% and 40% without adaptation in 2030 and 2090, and 22% and 60% with adaptation. In the southern Plains the gains are more modest, about 2% and 20% without adaptation in 2030 and 2090, and 5% and 35% with adaptation.

4.3. Results of the JGCRI Assessment

4.3.1. Agricultural production

The JGCRI study was an "integrated assessment" considering climate-change effects on crop production, water resources, unmanaged ecosystems, and irrigation. For crop-production effects this study relied on the EPIC model (version 7270) of Williams (1995) to simulate climate change impacts on agricultural production on the field scale for the entire conterminous USA. A more detailed explanation of the EPIC model is given here to foster understanding of the level of detail used in current agricultural assessments.

The EPIC model simulates climate change impacts on agricultural production on the field scale by calculating the maximum daily increase in plant biomass allowed by solar radiation incident on the field. The algorithms used to model potential plant growth are driven in EPIC by photosynthetically active radiation (PAR), the 0.4 – 0.7 micron portion of the solar spectrum. The amount of solar radiation captured by the crop is a function of leaf area index (LAI) and the amount converted into plant biomass is a function of the radiation use efficiency (RUE) which is crop-specific. Solar radiation also provides the energy that drives ET.

In EPIC the potential daily photosynthetic production of biomass is decreased by stresses caused by shortages of radiation, water and nutrients, by temperature extremes and by inadequate soil aeration. Each day's potential photosynthesis is decreased in proportion to the severity of the most severe stress of the day. As pointed out above, elevated atmospheric CO_2 concentration increases photosynthesis in C-3 plants and reduces ET in both C-3 and C-4 plants. Stockle et al.

(1992a, b) adapted EPIC to simulate the CO_2-fertilization effect on radiation use efficiency RUE and ET.[12]

Planting and harvesting dates in EPIC are based on accumulated heat units during the growing season and therefore, vary with different climate scenarios. Crop yields are estimated by multiplying aboveground biomass at maturity by a harvest index (proportion of the total biomass in the harvested organ).

The land areas used for the EPIC modeling in the JGCRI study were the 204 "4-digit" hydrologic unit areas of the lower 48 states (USGS 1987) shown in Figure 5-4. The GCM projections were downscaled by SCENGEN (see above) to fit these basins. BMRC and UIUC projections of climate change were made at global mean temperature increases of 1°C and 2.5°C. Crop behavior with and without the CO_2-fertilization effect (represented by CO_2 concentrations of 365 and 560 ppm) was also modeled. Model fields of corn, soybeans, and winter wheat were simulated for a representative farm in each of the 4-digit basins. Not all of these crops grow in all of these basins today, but climate change could conceivably alter the boundaries of the most productive regions of the country

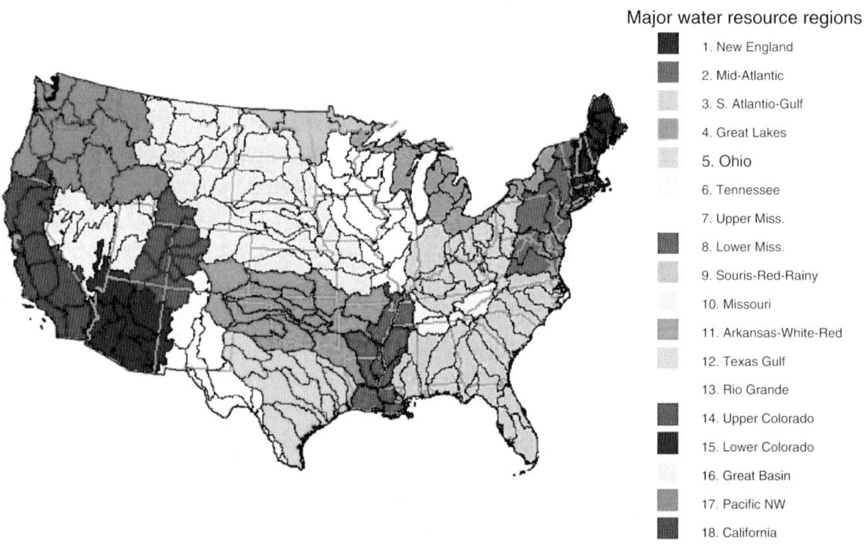

Figure 5-4. Major Water Resource Regions of the conterminous USA as defined by US Geological Survey (1987). The 204 modeling regions used in the JGCRI study are shown (*See Color Plates*)

[12] A recent review of studies conducted with Free Air Carbon Dioxide Enrichment (FACE) technology (Long et al. 2006) indicates that the CO_2-fertilization effect on both photosynthesis and evapotranspiration is less than had previously been estimated from the results of greenhouse, growth chamber and open-topped field chambers, the basis for the Stockle et al. (and many other) model algorithms. (*See Color Plates*)

for each crop. A "yield threshold" for change was defined for each crop by reference to their known core production areas (CPAs). The lowest EPIC simulated baseline (current climate) yield within the CPA was chosen as the yield threshold. These are: 2.5 Mg ha^{-1} for corn; 1 Mg ha^{-1} for soybean, and 1 Mg ha^{-1} for winter wheat. Yields lower than these identify areas where profitable production of these crops is unlikely under current or changed climates.

Figure 5-5 from Thomson et al. (2005a) shows results of these calculations for the conterminous USA. The brown colored areas are those in which production of the indicated crop is possible; the yellow hatching over brown demarcates the current core production areas. Green shows areas into which production may expand under the BMRC and UIUC GCM scenarios at GMT of 2.5°C. The ameliorating effects of CO_2-fertilization are not considered in this figure. Red indicates areas lost to production of the particular crop under these conditions. We focus here on results for the Great Plains. Under BMRC corn production becomes unprofitable in western portion of South Dakota, Nebraska, and eastern Wyoming. Under the UIUC scenarios, on the other hand, corn production becomes economically possible in northeastern Colorado, parts of Wyoming,

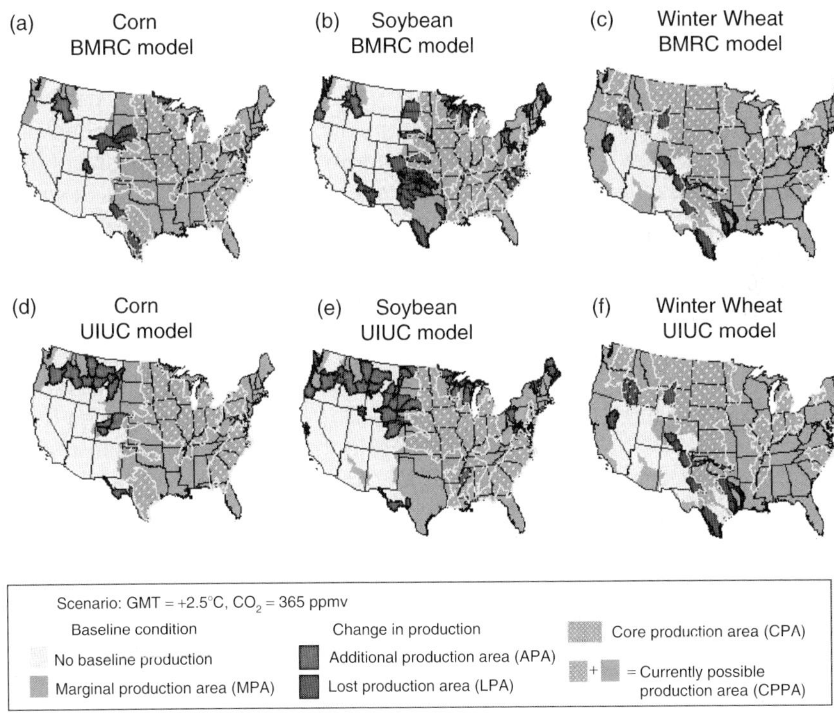

Figure 5-5. Regions projected to enter or leave production for three grain crops with the BMRC and UIUC GCMs at a global mean temperature increase of +2.5°C and CO_2 concentration of 365 ppmv (Source: Thomson et al. 2005a) (*See Color Plates*)

and much of Montana. While changes in the geography of corn production occur mostly on the western margins of the Plains, impacts on soybean production are more widespread. Under the dry BMRC scenario, potential production area is lost in the western corner of the North Dakota and South Dakota boundary, in southern South Dakota, central Kansas, Oklahoma, Texas, New Mexico, and southeast Colorado. New soybean areas appear in Montana, Wyoming, North Dakota, South Dakota, and Colorado under the moist UIUC scenario. The core production area for wheat also loses area to climate change at the western edge of the Plains under BMRC, but gains area under UIUC. Not surprisingly, the regions most affected by climate change, regardless of GMC scenario, are those on the margins of the regions in which they are currently grown.

4.3.2. Water resources

The hydrologic unit model of the USA (HUMUS) (Srinivasan et al. 1993) a geographic information system (GIS)-based modeling system, was used to simulate climate change effects on hydrology in the JGCRI study. HUMUS can be applied to a wide range of basin sizes depending on the availability of input data and the study objectives. In this study, the hydrologic cycle was simulated at the scale of the 8-digit USGS Hydrologic Unit Areas (USGS 1987), of which there are 2,101 in the conterminous US. In the modeling work climate, land use, and soil type are treated as uniform within each of these basins.

Water yield, a measure of net water flow out of each watershed, is calculated as the sum of surface and lateral flow from the soil profile and groundwater flow from the shallow aquifer. HUMUS and Soil and Water Assessment Tool (SWAT) runs on a daily time step with the same weather data inputs as described above for the EPIC model. The same algorithms are also used to account for the CO_2-fertilization effects on ET.

The HUMUS model was run for the BMRC and UIUC GCM projections scaled to each basin. Results were summed for each of the 18 major water resource regions of the USA shown in Figure 5-4. The Great Plains lie primarily within the Missouri, Arkansas-White-Red, and Rio Grande basins. Results for these basins in terms of water yield (most closely representing stream flows) are summarized in Table 5-1 (from Thomson et al. 2005b). This table shows that flows are reduced under BMRC in all three river basins at both levels of global warming (GMT=1.0°C and 2.5°C) and regardless of CO_2-fertilization. Under UIUC, on the other hand, flows increase in all three river basins. These results are consistent with the drying and wetting of these regions according to the BMRC and UIUC projections. Higher GMT worsens water yields still more under BMRC because of intensified drying and, conversely, increases water yields under UIUC wherein the greater global warming results in more precipitation for the Great Plains region. The higher CO_2 moderates losses under BMRC and increases water yields still further under UIUC.

Figure 5-6 shows the geographic distribution of the change in water yields. The general drying and wetting under BMRC and UIUC are evident. The southeastern corner of the Plains region shows the most severe drying under BMRC.

Table 5-1. Water yield at baseline climate and change from baseline (mm) for three Major Water Resource Regions under the BMRC and UIUC climate change scenarios (Thomson et al. 2005b, Table 1)

Scenario	GMT (°C)	CO_2 (ppm)	Missouri	Arkansas-White-Red	Rio Grande
Baseline	0	365	107	235	40
BMRC	1.0	365	−16	−31	−10
		560	−11	−19	−8
	2.5	365	−35	−73	−21
		560	−31	−62	−21
UIUC	1.0	365	20	50	15
		560	26	65	18
	2.5	365	55	139	46
		560	63	157	51

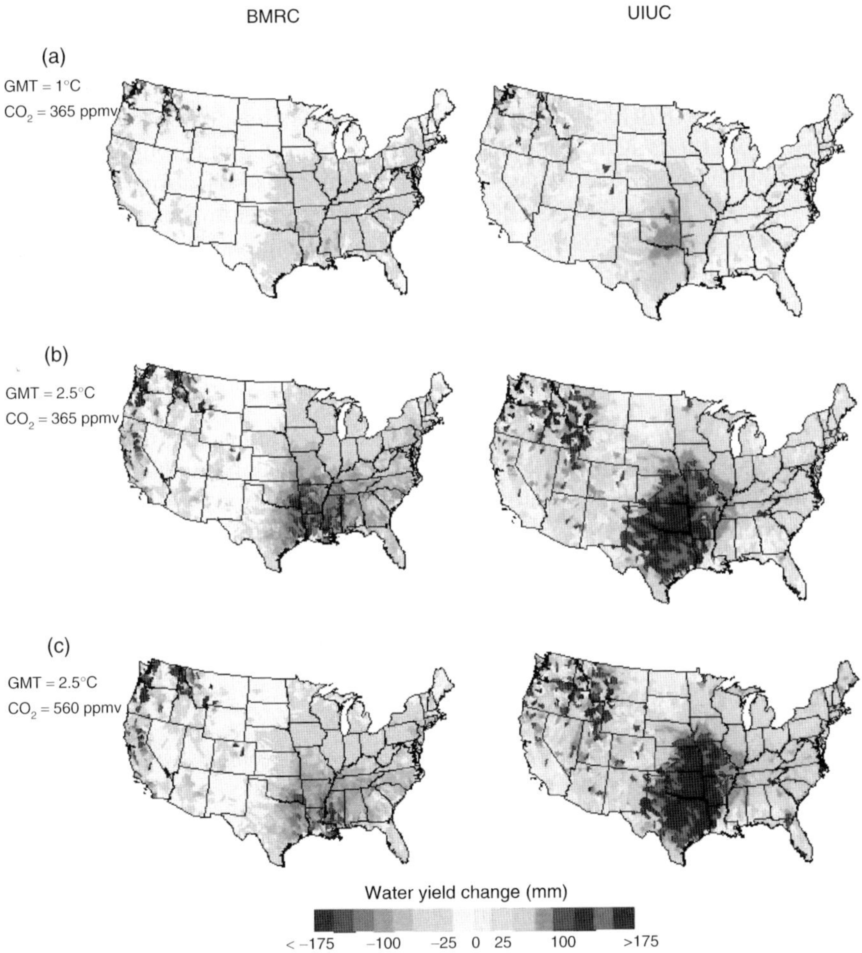

Figure 5-6. Water yield change from baseline (mm) for two GCMs with increasing global mean temperature (GMT) with and without the CO_2-fertilization effect (Thomson et al. 2005b)(*See Color Plates*)

Almost all of the USA shows increased water yields under UIUC, but increases are greatest in the southern Plains. Notable increases in water yield of 75% or more extend into eastern Kansas and Nebraska as well.

4.3.3. Implications for irrigation

Under the bountiful (indeed, in some regions excessive) precipitation of the UIUC scenario, the need for irrigation in the Great Plains is diminished, as it is in the country as a whole (Thomson et al 2005c). Under the BMRC scenarios, on the other hand, irrigation declines because of lost water yields. In such a case agriculture may be doubly disadvantaged as sectors other than agriculture will compete for shares of the diminished water supplies. Conflicts among competing users of water, already a fact of life on the Plains, would surely intensify.

If the past is any guide to the future, we can expect that farmers on the Plains will find ways, if not to fully adapt to a worsened climate, at least to partially offset its impacts in ways that reduce the need for irrigation of water. These include development of drought resistant or tolerant cultivars of the current crops, and the introduction of less water intensive new crops. Efforts will be made to increase irrigation efficiency through increased use of water conserving application methods such as are described in Chapter 4, and through the use of monitoring devices that determine optimum timing and application rates. Nonetheless, should warming and drying occur in the Plains region—more likely, most specialists agree, than wetting—reliance on irrigation to maintain its productive capacity will be more difficult.

5. SUMMARY: POSSIBLE IMPACTS OF CLIMATE CHANGE ON GREAT PLAINS AGRICULTURE

From studies such as those described above and from other major information assembly efforts such as IPCC, a generally good understanding has emerged of what the impacts of climate change could be for agriculture. The IPCC process turns out comprehensive reports on climate change science, impacts and mitigation strategies every 5 years. The IPCC Third Assessment Report (TAR) covered the period 1995 – 2000 and was published in 2001. The Fourth Assessment Report is to be published in 2007. The TAR sections on agriculture and water resources and its report on North America provide useful, but very general, assessments of how its agriculture may be affected by climate change. A review of all these sources leads this writer to conclude that:

- The CO_2-fertilization effect should raise crop yields, but not by the amount demonstrated in controlled experimental environments, because in the field the effects of variable climate, soil fertility, and tilth are not always optimum and crops face competition from weeds, some of which may benefit as much or more from elevated CO_2, and from stress due to insects and diseases. The benefits of CO_2-enrichment in increased water use efficiency may be offset by higher temperatures and increased evaporative demand (see footnote 12).

- In some regions (the Great Plains is a candidate) climatic changes may lead to increased leaching of nutrients, loss of soil organic matter, and increased soil erosion, but such effects can be combated or alleviated by crop rotations, conservation tillage and improved nutrient management.
- Livestock can also be affected both directly and indirectly. Livestock are stressed by warmer, drier conditions although warmer winters may improve survival rates on the open range. Warmer, drier conditions—more likely than not on the Plains according to the majority of GCMs—may reduce forage quality and water availability for the livestock. Since livestock can be moved if necessary and feed can be imported, impacts on the livestock sector may be delayed or offset, but ultimately livestock production would become untenable where producers cannot count on reliable feed supplies.
- Modest improvements in crop growing conditions are projected by several GCMs for the northern tier of the American states and for the Canadian Prairie Provinces. Warming and a consequent lengthening of the growing season could encourage corn production in North Dakota and a further northward advance of winter wheat in Canada.
- Whether small or large, and whether accompanied by precipitation change or not, temperature increases have certain consistent effects on crop growth and yield. Annual crops can be planted sooner in northern climes with earlier thawing of the ground in spring. As they are for the most part driven by thermal-time (i.e., heat units), the passage of the plant's phenological (developmental) stages is accelerated and all occur earlier in the growing season.
- Two disadvantages accompany warming. During the reproductive phase, certain crops (notably corn) are very sensitive to high temperatures. Pollination can be unsuccessful in such a case with the result of deep reductions in yield. Another disadvantage is the earlier achievement of crop maturity. With a shortened growing season, there is less time for the plants to synthesize and store starch or other products in the seed or tuber. So, without development of new cultivars better suited to the greater warmth, crop yields could suffer from a warmer climate.
- On the other hand, frosts are likely to be less frequent and severe as the climate warms, so the risk of early-season and late-season damage is decreased. This truism may not be totally reliable, however, especially in spring. Neild et al. (1979) have shown that as a warmer spring advances the development of newly planted crops, they may actually be at a more vulnerable phenological stage when frosts do occur. On the other hand, perennial crops, especially those grown for forage, hay, or biomass should benefit from warming because of a lengthened growing season. General warming will also permit the growth of major crops such as corn in higher latitude locations than at present. But the importation of cultivars from lower latitudes is not always simple because they are generally adapted to the length of day where they are grown.
- Higher temperature has another important influence on crop growth through its effects on water use. Evapotranspiration is a function of leaf and air temperatures.

Warm air provides energy to evaporate water. The dryness of the air in contact with the crop's leaves is also a function of air temperature. The drying power of the air (its vapor pressure deficit) increases exponentially with temperature. Relative humidity (the ratio of actual water vapor pressure to the saturation vapor pressure at the ambient air temperature) falls rapidly with rising temperature. So, whether there is more precipitation or less, the actual crop water use (if water is present) increases with warming.

- The impacts of change in precipitation are intuitive: too much rain leads to flooding or at least to saturated soil, anaerobic conditions, and leaching of nutrients; too little rain leads to moisture stress on plants, wilting, closure of the leaf stomata (pores) so that photosynthesis is shut down. If severe and protracted, lack of precipitation means drought and, ultimately, crop failure.

With these general concepts in mind and recognizing that, however uncertain the specific climatic changes awaiting the Great Plains, profound impacts are possible in the region's agriculture as this century progresses. We next explore the possibility that the Great Plains region may have a role to play in mitigating or at least slowing the progress of global climate change, perhaps to its own long-term benefit.

REFERENCES

Blasing, T.J. and Solomon, A.M., 1983, *Response of the North American Corn Belt to Climatic Warming*, Carbon Dioxide Research Division, US Department of Energy Report DoE/NBB-004.

Blasing, T.J. and Jones, S., 2005, *Current Greenhouse Gas Concentrations,* Report of the Carbon Dioxide Information Analysis Center, Oak Ridge National Laboratory, US Department of Energy, Oak Ridge, Tennessee, http://cdiac.ornl.gov/pns/current_ghg.html

Brown, R.A., Rosenberg, N.J., Hays, C.J., Easterling, W.E., and Mearns, L.O., 2000, Potential production and environmental effects of switchgrass and traditional crops under current and greenhouse-altered climate in the Central United States: a simulation study, *Agric. Ecosyst. and Environ.*, **78**:31–47.

Flato, G.M., Boer, G.J., Lee, W.G., McFarlane, N.A., Ramsden, D., Reader, M.C., and Weaver, A.J., 2000, The Canadian Centre for Climate Modelling and Analysis: global coupled model and its climate, *Clim. Dyn.* **16**:451–467.

Georgi, F. and Mearns, L.O., 1991, Approaches to the simulation of regional change: a review, *Rev. Geophys.* **29**: 191–216.

Georgi, F., Mearns, L.O., Sheilds, C., and McDaniel, L., 1998, Regional nested model simulations of present day and 2 × CO_2 climate over the central plains of the US, *Clim. Chang.* **40**:457–493.

Glantz, M.H. and Ausubel, J.H., 1984, The Ogallala aquifer and carbon dioxide: comparison and convergence, *Environ. Conser.* **11**:123–131.

Hulme, M., Jiang, T., and Wigley, T.M.L., 1995, *SCENGEN: A Climate Change Scenario Generator: User Manual,* Climatic Research Unit, University of East Anglia, Norwich, UK, 38 pp.

Intergovernmental Panel on Climate Change (IPCC), 2001a, *Third Assessment Report—Climate Change 2001. Report of Working Group I. The Scientific Basis,* Chap. 2: Observed climate variability and change, Cambridge University Press, Cambridge and New York, pp. 99–181.

Intergovernmental Panel on Climate Change (IPCC), 2001b, *Third Assessment Report—Climate Change 2001. Report of Working Group II. Impacts, Adaptation and Vulnerability*, Chap. 15: The regional impacts of climate change, North America, Cambridge University Press, Cambridge and New York, pp. 735–800.

Johns, T. C., Carnell, R. E,. Crossley, J. F., Gregory, J.M., Mitchell, J. F. B., Senior, C.A., Tett, S. F. B., and Wood, R. A., 1997, The Second Hadley Centre coupled ocean-atmosphere GCM: model description, spin-up and validation, *Clim. Dyn.* **13**:103–134.

Karl, T.R., Knight, R.W., Gallo, K.P., Peterson, T.C., Jones, P.D., Kukla, G., Plummer, N., Razuvayev, V., Lindseay, J., and Charlson, R.J., 1993, A new perspective on recent global warming: asymmetric trends of daily maximum and minimum temperature. *Bull. Amer. Meteorol. Soc.* **74**:1007–1023.

Karl, T.R., Knight, R.W., Easterling, D.R., and Quayle, R.G., 1996, Indices of climate change for the United States, *Bull. Amer. Meteorol. Soc.* **77**:279–292.

Kiehl, J.T. and Trenberth, K.E., 1997, Earth's annual global mean energy budget, *Bull. Amer. Meteorol. Soc.* **78**:197–208.

Long, S.P., Ainsworth, E.A., Leakey, A.D.B., Nosberger, J., and Ort, D.R., 2006, Food for thought. Lower than expected crop yield stimulation with rising carbon dioxide concentrations. Science, **312**, 1918–1921.

Marland, G., Boden, T.A., and Andres, R.J., 2005, *Global, Regional, and National CO_2 Emissions*, in *Trends: A Compendium of Data on Global Change*, Carbon Dioxide Information Analysis Center, Oak Ridge National Laboratory, US Department of Energy, Oak Ridge, Tennessee.

McAveney, B.J., Colman, R., Fraser, J.F., and Dhani, R.R., 1991, *The Response of the BMRC AGCM to a Doubling of CO_2*, Bureau of Meteorology Research Centre (BMRC) Technical Memorandum No.3, Melbourne, Australia.

National Academy of Sciences, 1992, *Policy Implications of Greenhouse Warming: Mitigation, Adaptation and the Science Base,* NAS Press, Washington, DC, pp. 541–566.

Neild, R.E., Richman, H.N., and Seeley, M.W., 1979, Impacts of different types of temperature change on the growing season of maize, *Agric. Meteorol.* **20**: 367–374.

Newman, J. E., 1980, Climate change impacts on the growing season of the North American Corn Belt, *Biometeorology*, **7**: 128–142.

Parmesan, C. and Yohe, G., 2003, A globally coherent fingerprint of climate change impacts across natural systems, *Nature*, **421**:37–42.

Reilly, J., Tubiello, F., McCarl, B., Abler, D., Darwin, R., Fuglie, K., Hollinger, S., Izaurralde, R.C., Jagtap, S., Jones, J., Mearns, L., Ojima, D., Paul, E., Paustian, K., Riha, S., Rosenberg, N.J., and Rosenzweig, C., 2003, US climate and agriculture: new results, *Clim. Chang.* **57**: 43–67.

Root, T.L., Price, J.T., Hall, K.R., Schneider, S.H., Rosenzweig, C., and Pounds, J. A., 2003, Fingerprints of global warming on wild animals and plants, *Nature*, **421**:57–60.

Rosenberg, N.J., 1982, The increasing CO_2 concentration in the atmosphere and its implication on agricultural productivity. II. Effects through CO_2 induced climate change, *Clim. Chang.* **4**:239–254.

Rosenberg, N.J., 1992, Adaptation of agriculture to climatic change, *Clim. Chang.* **21**:385–405.

Rosenberg, N.J. (ed.), 1993, Towards an Integrated Assessment of Climate Change: the MINK study, *Clim. Chang.* **24**:1–173.

Rosenberg, N.J. and Edmonds, J.A. (eds.), 2005, Climate change impacts for the conterminous USA: an integrated assessment, *Clim. Chang.* **69**:1–162.

Schlesinger, M.E., 1997, *Geographical Scenarios of Greenhouse-Gas and Anthropogenic-Sulfate-Aerosol Induced Climate Changes,* report for Energy Modeling Forum (EMF-14), University of Illinois, Urbana–Champagne, 85 pp.

Smith, S.J., Thomson, A.M., Rosenberg, N.J., Izaurralde, R.C., Brown, R.A., and Wigley, T.M.L., 2005, Climate change impacts for the conterminous USA: an integrated assessment, Part 1. Scenarios and context, *Clim. Chang.* **69**:7–25.

Srinivasan, R., Arnold, J.G., Muttiah, R.S., Walker, C., and Dyke, P.T., 1993, Hydrologic Unit Model for the United States (HUMUS), in *Advances in Hydroscience and Engineering*, CCHE, School of Engineering, University of Mississippi, Oxford, Mississippi.

Stockle, C.O., Williams, J.R., Rosenberg, N.J., and Jones, C.A., 1992a, A method for estimating direct and climatic effects of rising atmospheric carbon dioxide on growth and yield of crops: Part I. Modification of the EPIC model for climate change analysis, *Agric. System*, **38**:225–238.

Stockle, C.O., Dyke, P.T., Williams, J.R., Jones, C.A., and Rosenberg, N.J., 1992b, A method for estimating direct and climatic effects of rising atmospheric carbon dioxide on growth and yield of crops: Part II. Sensitivity analysis at three sites in the Midwestern USA, *Agric. Systems*, **38**:239–256.

Thompson, L.M., 1986, Climatic change, weather variability, and corn production, *Agron. J.* **78**:649–653.

Thomson, A.M., Brown, R.A., Rosenberg, N.J., Izaurralde, R.C., and Benson, V., 2005a, Climate change impacts for the conterminous USA: an integrated assessment. Part 3. Dryland production of grain and forage crops, *Clim. Chang.* **69**:43–65.

Thomson, A.M., Brown, R.A., Rosenberg, N.J., Srinivasan, R., and Izaurralde, R.C., 2005b, Climate change impacts for the conterminous USA: an integrated assessment. Part 4. Water resources, *Clim. Chang.* **69**:67–88.

Thomson, A.M., Rosenberg, N.J., Izaurralde, R.C., and Brown, R.A., 2005c, Climate change impacts for the conterminous USA: an integrated assessment. Part 5. Irrigated agriculture and national grain production, *Clim. Chang.* **69**:89–105.

United States Geological Survey, 1987, *Hydrologic Unit Maps*, US Government Printing Office, Washington, DC.

United States Global Change Research Program (USGCRP), 2002, *US National Assessment of the Potential Consequences of Climate Variability and Change Sector: Agriculture*. Reilly, J. (ed.), Cambridge University Press, Cambridge and New York.

Williams, J.R., 1995, The EPIC model, in Singh, V.P. (ed.), *Computer Models in Watershed Hydrology*, Water Resources Publications, Highlands Ranch, Colorado, pp. 909–1000.

CHAPTER 6

A ROLE FOR THE PLAINS IN COMBATING CLIMATE CHANGE

1. INTRODUCTION

It may, as the previous chapter explores, be necessary on the Great Plains, as elsewhere, to adapt to climatic change. But adaptation becomes more and more difficult as the climate changes become more profound. And, unlike the effects of poor land management (erosion, water contamination), the causes of climate change are not purely local, but stem from fossil fuel emissions and land use changes occurring all over the world, the Great Plains not excepted. Although the Plains are vulnerable to the effects of climate change, the region may have a role—perhaps an important role—to play in mitigating climate change.

If climate change is to be controlled, mitigation strategies must be deployed worldwide. To the extent that land use change contributes to global warming, the simple, obvious (although not easy) solution is to alter such practices as deforestation that lead to CO_2 and CH_4 emissions. But where fossil fuel combustion and other activities that emit CO_2 and the other greenhouse gases (GHGs) are involved, we must look for ways to reduce the need for these fuels or improve the efficiency with which they are used. And for the CO_2 emissions that continue, it may be necessary to "grab and bag" them at the smokestack and maybe eventually at the tailpipe too and somehow store these gasses away to avoid their reaching the atmosphere.

Increasing energy end-use efficiency is an obvious strategy. Many improvements have already been made in water heaters, furnaces, refrigerators, and other types of appliances to increase their energy efficiency. In the realm of transportation, internal combustion engines have been improved and hybrid internal combustion-electric battery engines (as in Toyota's *Prius*) have been introduced to the market. The plague of the "gas-guzzling" sport utility vehicles continues to nullify the overall effect of these improvements. Hydrogen fuel cells also offer an alternative to the CO_2-emitting internal combustion engine.

The need for fossil fuels to generate electric power can be reduced by increasing energy conversion efficiencies through the use of gas turbines, coal gasification, and other means. The US Department of Energy's (USDOE) Fossil Energy Division aims by 2010, to develop advanced power systems capable of achieving between 45% and 50% electrical efficiency (typically in the 30% range with current technologies) at a capital cost of $1,000 per kilowatt or less for a coal-based power plant.[1]

More nuclear power plants, more hydroelectric plants, more wind turbines, more passive and active solar energy facilities—all are proposed as partial substitutes for fossil fuel-driven generation facilities. And, last but not least, biomass which recycles atmospheric CO_2, rather than adding more of it to the atmosphere, is rapidly gaining credibility and importance.

The "grab it and bag it" approach to reducing the negative effects of CO_2 emissions is to strip that substance from the efflux of gases at the smokestack and to remove it from circulation by depositing it in deep oceanic or terrestrial geological sinks. These include depleted oil and gas fields, abandoned coal mines and saline aquifers.

But all of the technological improvements or innovations listed above carry with them certain disadvantages and/or associated, although not necessarily insurmountable, risks that must be factored into the overall strategy for CO_2 emissions abatement. The nuclear option, for example, is unpopular in the USA and Scandinavia because of fear that events like Chernobyl[2] and Three Mile Island[3] could again occur. The dangers associated with radioactive wastes and the scientific, engineering, and political difficulties of finding permanent storage for them contribute to this aversion. At this writing and after decades of research, planning, and preliminary construction, the Yucca Mountain repository in Nevada[4] is yet to accept any nuclear power plant waste for permanent storage.

Hydroelectric power generation emits no GHGs (except through the machinery and materials, particularly cement, used in their construction and maintenance) but there are few remaining sites in the USA for large new hydroelectric facilities.

---

[1] http://www.fe.doe.gov/programs/powersystems/gasification/index.html

[2] The Chernobyl (Ukraine) reactor accident occurred in 1986, and is attributed to a flawed reactor design and operation by inadequately trained personnel. The resulting steam explosion and fire released at least 5% of the radioactive reactor core into the atmosphere and downwind. Source: World Nuclear Association, 114 Knightsbridge, London SW1X 7LJ. (http://www.uic.com.au/nip22)

[3] The accident at the Three Mile Island Unit 2 (TMI-2) nuclear power plant near Middletown, Pennsylvania, on March 28, 1979, was the most serious in US commercial nuclear power plant operating history. Although it led to no deaths or injuries to plant workers or members of the nearby community, it brought about sweeping changes involving emergency response planning, reactor operator training, human factors engineering, radiation protection, and many other areas of nuclear power plant operations. Source: http://www.nrc.gov/reading-rm/doc-collections/fact-sheets/3mile-isle.html.

[4] Yucca Mountain, a remote federally protected desert site in Nye County, Nevada, has been under study by the US Department of Energy since 1978 as a long-term geologic repository for spent nuclear fuel and high-level radioactive waste stemming from nuclear power generation and national defense programs. It is ~160 km (100 miles) northwest of Las Vegas, Nevada. The Department of Energy is at this writing in the process of preparing an application to obtain the Nuclear Regulatory Commission license to proceed with construction of the repository. (http://www.ocrwm.doe.gov/ymp/index.shtml)

Indeed, the environmental impacts of hydrofacilities—the inundation of fertile agricultural valleys, interference with reproductive cycle of salmon and other anadromous species, and other ecological concerns make it unlikely that any more Hoover or Grand Coulee dams will be constructed in the USA in the coming years. In fact, the environmental community has been influential in bringing about the removal of small dams and restoration of natural flows in rivers from New England[5] to the Olympic Peninsula[6] of Washington State.

Canada, on the other hand, has an estimated hydroelectric potential of more than 182,000 megawatts (MW). According to a Canadian government website,[7] taking technical, environmental, and economic factors into account, over 34,000 MW of this potential may be practical for future development by electric utilities.

The USA has many areas with abundant winds, particularly in the Midwest and the Great Plains. Engineering has greatly increased the efficiency of wind turbines in recent decades. Driven by the high costs of energy, installation of turbines in the USA reached its then highest level in 2005. The wind power industry has had its ups and downs over the past decade but industry outlook appears to be bright. According to a recent industry report[8]

> North American wind power is expected to see a more than fourfold increase in wind power plants in operation by 2010. The U.S. is expected to grow from just over 6,700 MW to over 28,000 MW by 2010. Starting from a lower base of nearly 450 MW in 2004, Canada's wind power base will grow even more quickly to over 6,200 MW by 2010.

While some (this writer included) find the view of a field of tower-mounted turbines with gracefully spinning windfoils aesthetically pleasing, others denounce them as a source of visual and noise pollution and for dangers they pose to bird populations and, hence, to local ecosystems. Habitat destruction is of particular concern where turbines are to be placed in remote mountainous locations as forestland must be cleared, roads built, and transmission lines installed. A distinct disadvantage of the wind turbine is that, when the wind does not blow, backup power must be brought in through the grid from other, often fossil fuel-powered, sources.

---

[5] On July 1, 1999, the 160-year-old Edwards Dam in Augusta, Maine, was removed allowing the Kennebec River to flow freely from Waterville to the sea. The work is the result of a decades-long effort to restore Maine fisheries and return the Kennebec to its free-flowing status. (http://www.state.me.us/spo/sp/edwards/)

[6] The Elwha River drains the Olympic Mountains of Washington, flowing northward to the Strait of Juan de Fuca. Construction of two hydroelectric dams in the early 1900s resulted in the loss of ~95% of the anadromous salmon spawning habitat on the river. The Elwha River Ecosystem and Fisheries Restoration Act, enacted by Congress In 1992, authorized removal of the dams in order to restore the once-plentiful salmon runs in the river. Dam removal is currently slated to begin in early 2008. (http://soundwaves.usgs.gov/2005/02/research.html)

[7] http://www.canren.gc.ca/tech_appl/index.asp?CaId=4&PgId=26)

[8] Wind Power 2005 in Review, Outlook for 2006 and Beyond. (http://www.renewableenergyaccess.com/rea/news/story?id=41304)

Solar power facilities are also not universally popular. This non-CO_2-emitting source requires the construction of complex systems that demand large areas of land in order to provide surface for capture of significant quantities of solar energy. The logical place for such facilities is in desert areas where land is relatively cheap and most days are sunny. But solar energy is converted to electrical energy only when the sun shines. So other sources of power are needed during night time and on cloudy days, or the energy collected when the sun shines must somehow be stored for use when it does not.

The foregoing discussion leads us to biomass which recycles CO_2 from the atmosphere rather than adding more to it and can substitute for fossil fuel. Biomass is the prime focus of this chapter; its benefits and disadvantages are discussed in detail in subsequent sections. And, for the purposes of this book, sequestration of carbon in agricultural soils, also discussed at length below, will be considered under the rubric of biomass.

2. TECHNOLOGICAL FIXES

2.1. Background

The Joint Global Change Research Institute, a collaboration activity of the USDOE's Pacific Northwest National Laboratory and the University of Maryland—College Park, operates a Global Technology Strategy Program (GTSP). Economists, engineers, agronomists, and other natural scientists in this group have developed a number of ways to estimate the potential role of each of the mitigation strategies discussed above in closing the gap between tolerable annual fossil fuels emissions to the atmosphere and those likely to occur given expected energy usage, the latter a function of anticipated growth in population, rising standards of living, and general economic change in the future.

The result of one such analysis is shown in Figure 6-1a from Edmonds and Clarke (2004). The top line in this figure shows anticipated emissions over the course of this century assuming no change in technology from 1990. The bottom line shows the emissions permissible if the atmospheric concentration of CO_2 is constrained to no more than 550 ppmv. As noted in Chapter 5 that concentration had already risen to 379 ppmv by the end of 2004.

The IPCC has developed many scenarios of greenhouse gas emissions (GHGEs) for the 21st century where the variables are estimated population, standards of living, technological advances, and implementation of emission control strategies. One of the many scenarios developed, labeled IS92a, has been widely used, since it is fairly close to the mean of those available in the course and direction of its projections. IS92a assumes that "business as usual" prevails—meaning that emissions-relevant technologies improve with time at a rate consistent with past experience. Even with continuous progress in energy production and in end-use efficiency, the IS92a scenario projects that the emission of carbon to the atmosphere (in the form of CO_2) will increase from ~7+ GtC per annum in 2005 to ~20 GtC per annum in 2095. As discussed in Chapter 5, CO_2 would rise from its current atmospheric concentration of ~380 to over 700 ppmv under such a carbon load (Figure 6-1b). Were there to be no improvement from 1990s technology, both the carbon emissions and the atmospheric CO_2 concentration are literally "off the charts."

The bottom line in Figure 6-1a indicates the course of carbon emissions that would be required to stabilize the CO_2 concentration at 550 ppmv, assumed but not proven to be consistent with the goal of the United Nations Framework Convention on Climate Change (United Nations 1992) "to avoid dangerous interference with the climate system." This line results from the work of Wigley, Richels, and Edmonds (1996, hereafter WRE) who investigated the time course of CO_2 emissions required to stabilize it in the atmosphere at concentrations ranging from 350 to 1,000 ppmv.

Returning to Figure 6-1a: the area between the IS92a (BAU) projections (middle line) and the indicated path to stability at 550 ppmv (bottom line) represents the total amount by which carbon emissions must be decreased beginning in around 2020 and projecting to the end of this century.

The top line in Figure 6-1a represents carbon emissions over the century as they would be assuming no change in technology from 1990. This is not a very likely assumption and is intended by Edmonds and Clarke (2004) to demonstrate how great an effort in terms of conservation, energy production efficiencies, carbon sequestration, and new, nonfossil fuel technologies will be needed just to

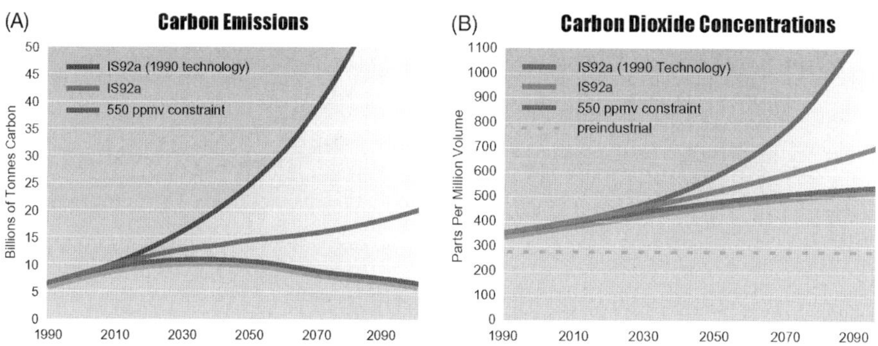

Figure 6-1. Comparison of IPCC IS92a "business as usual" scenario (*middle line*), the same scenario without assumed improvements in energy technology and an emissions path that constrains the concentration of CO_2 in the atmosphere to 550 ppmv. The left panel plots emissions of CO_2 and the right panel plots associated CO_2 concentrations in the atmosphere (Source: Edmonds and Clarke 2004)

assure that carbon emissions do not exceed 20 GtC per annum and a consequent rise of atmospheric CO_2 concentration to more than 700 ppmv.

As Edmonds and Clarke explain, the concentration of CO_2 is associated with cumulative rather than annual emissions. Therefore, all emission paths that stabilize its concentration have a pattern of emissions that peak and then decline monotonically thereafter until, eventually, emissions approach zero. Figure 6-2, also from Edmonds and Clarke (2004), shows that the lower the permissible CO_2 concentration, the smaller the annual emissions allowable and the sooner they must peak and begin their decline.

2.2. Closing the gap between likely and desirable emissions and CO_2 concentrations

Options for mitigating carbon emissions have been described above. But how much can each of these technological options really contribute to closing the gap between allowable and expected emissions? Considerable effort has been expended by physicists, engineers, and economists to answer these questions. Although a review of the extensive literature on this subject is beyond the scope of this book, some syntheses illustrate the possible market penetration and impact of technologies that may close the gap between projected and allowable carbon emissions. All of the technologies proposed are worthy of close exami-

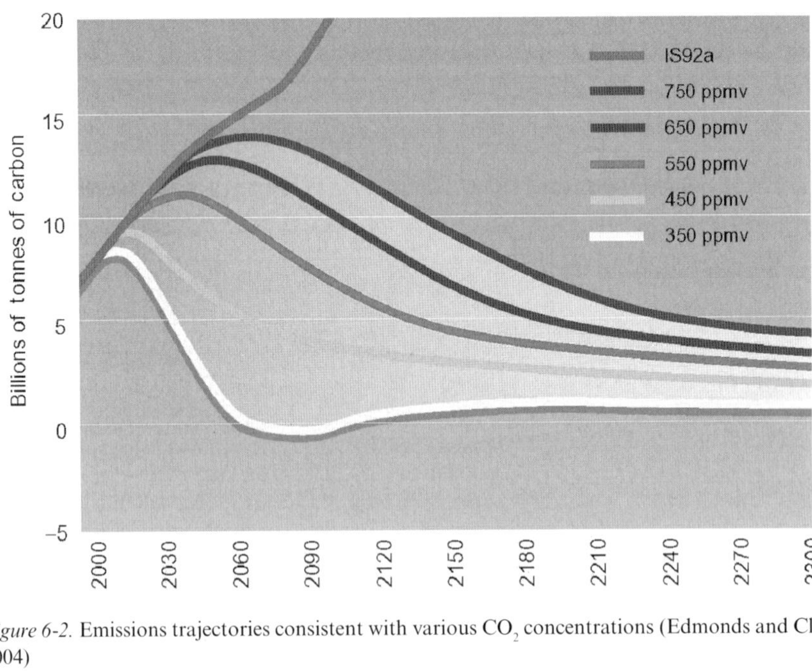

Figure 6-2. Emissions trajectories consistent with various CO_2 concentrations (Edmonds and Clarke 2004)

nation. Again, however, the emphasis in this chapter is on the potential role of biomass.

The results of one comprehensive analysis in which the range of mitigation technologies is assessed are shown in Figure 6-3 from Edmonds and Clarke (2004) based on their MiniCAM integrated assessment model (Edmonds et al. 1994). The anticipated growth of global energy demand and the contributions of various energy sources to meeting this demand under a set of constraints imposed by the IPCC SRES B2 scenario are shown in this figure.[9] The upper left figure shows global energy demand rising from 400 EJ (exojoule = 10^{18} joules) in

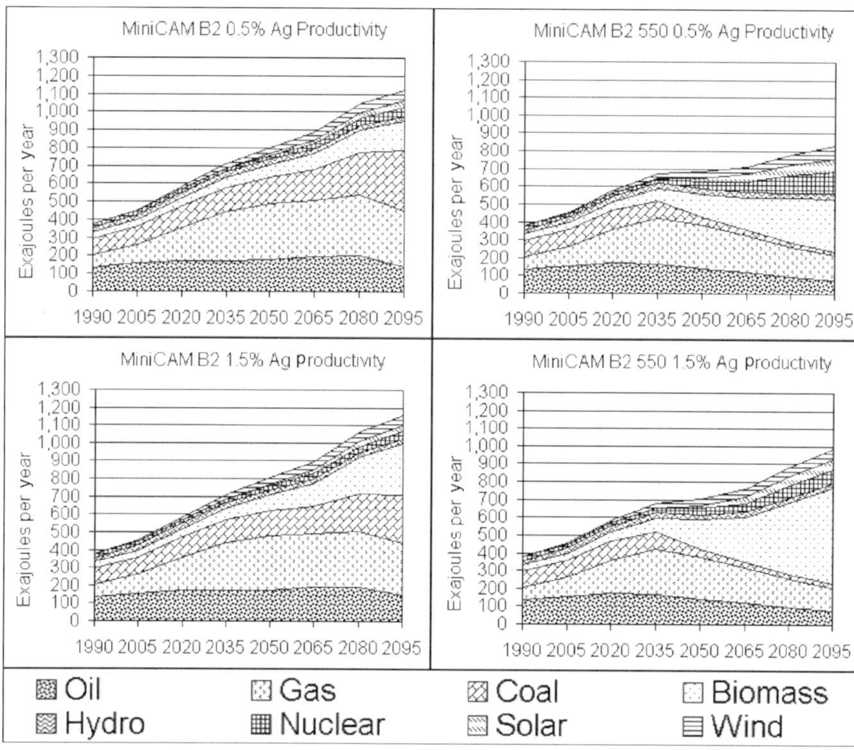

Figure 6-3. Commercial biomass energy in the global energy system at two rates of increasing agricultural productivity, with and without limitations on cumulative carbon emissions to avoid exceedence of a 550 ppmv CO_2 atmospheric concentration. (Carbon capture and disposal not considered in these calculations.) (Source: Edmonds and Clarke 2004)

[9] As compared with IPCC IS92a scenario the B2 scenario envisions slower growth in population with a more rapidly evolving economy and more emphasis on environmental protection and, therefore, produces lower emissions and less global warming. Source: IPCC, 2000. Special Report on Emission Scenarios.

1990 to 1,100 EJ in 2090, assuming no limitation is imposed on the atmospheric concentration of CO_2. Oil, gas, and coal continue to provide the bulk of the world's energy needs. A growing, but still small, role is given to biomass on the assumption that its productivity increases at the very modest rate of 0.5% per annum.

The lower left figure foresees a larger role for biomass in meeting the unconstrained 1,100 EJ global energy demand, because agricultural productivity increases at the rate of 1.5% per annum which is actually lower than the historical rate of crop productivity growth during the latter half of the 20th century. This increased productivity reduces per tonne production costs, making biomass more economically competitive as a source of energy.

The figures on the right are based on the assumption that global energy demand is reduced in an attempt to limit atmospheric CO_2 concentration to no more than 550 ppmv. In such a case, use of fossil fuels declines sharply and coal, the most carbon-intensive of the fossil fuels, is phased out entirely. The contribution of biomass grows significantly (upper right figure) in this scenario to offset reduction in fossil fuel use despite the fact that, as before, the productivity of biomass (actually all agriculture) continues to increase at the sluggish rate of 0.5% per annum.

In the lower right-hand figure biomass becomes the dominant source of energy despite the greater demand for energy shown in the upper right figure (1,000 vs 800 EJ). The contribution of biomass by the end of the century is about equal to that of all other energy sources combined. This is the consequence of the higher rate of productivity gain (1.5% per annum) in combination with the 550 ppmv atmospheric CO_2 concentration constraint. It is interesting to note that Edmonds and Clark foresee much less change in the contribution that wind and solar energy make under the B2 scenario.

Pacala and Socolow (2004) use the concept of "wedges" in describing a similar approach to evaluating opportunities for closing the gap between anticipated carbon emissions and that needed to maintain an acceptable atmospheric CO_2 concentration. Each wedge represents a technology that by 2054 reduces emissions by at least 1 GtC per annum or by 25 GtC from 2004 to 2054. One of these wedges is biomass fuel substitution for fossil fuel. This, they calculate, would involve production of 34 million barrels per day of ethanol, requiring production of high yielding biomass on a global land area of 250 million hectares.

3. SOIL CARBON SEQUESTRATION

3.1. Background

Sequestration of carbon in agricultural, range and forest soils is one biologically based mitigation strategy. The conversion of native forests, wetlands, and grasslands to agriculture resulted in the oxidation of organic matter in soil and the release of carbon to the atmosphere in the form of CO_2. It is thought that 55 GtC may have been released from soils worldwide since the mid-19th century (Paustian et al. 1998). Conventional tillage involving annual plowing, disking, and harrowing continually exposes soil organic matter (SOM) and fresh plant residues to oxidation and accelerates their decomposition. Minimum-till and no-till agricultural systems facilitate the restoration of organic matter in soils toward its preconversion levels—a form of "carbon sequestration." It is estimated by Cole et al. (1996) in the IPCC Second Assessment Report that from 40 to 80 GtC could potentially be restored to agricultural soils over the course of this century. This would represent a restoration of two-thirds of the carbon lost during land conversions.[10] The special role of this form of carbon sequestration is shown in Figure 6-4.

This figure shows results of a MiniCAM (Mini Climate Change Assessment Model, Edmonds et al. 1994) analysis, assuming the full potential (80 GtC) of soil carbon sequestration can be realized in croplands. With that being so, the gap between the IS92a "business as usual" emissions pattern and the WRE 550 ppmv pathway could be bridged by soil carbon sequestration alone, buying time—perhaps 35 years—for other mitigation options to gain market share and exert their influence. After around 2035 in this figure, further emissions reductions must come from changes in the energy system including alternative energy sources (fuel mix) and conservation (energy intensity).

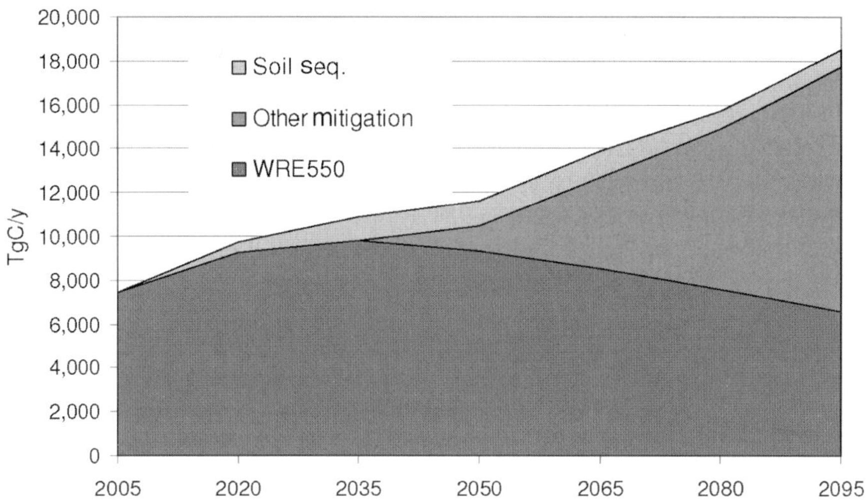

Figure 6-4. Role of soil carbon sequestration (SCS) in mitigating greenhouse warming. *Top line*, global carbon emissions projected by the IPCC "business as usual" scenario from 2005 to 2095. *Bottom line*, WRE (Wigley et al. 1996) emissions path required to stabilize the atmosphere at a CO2 concentration of 550 ppmv. Until about 2035 in this MiniCam model (Edmonds et al. 1994) projection, soil carbon sequestration alone could allow the constraint to be met. The contribution of soil carbon sequestration diminishes in importance after 2035 as various conservation and fossil fuel substitution technologies are deployed (Rosenberg et al. 1999)

---

[10] In this calculation a global average agricultural sequestration rate of 0.4–0.8 GtC occurring over a period of 50–100 years is assumed. Both 0.4 GtC × 100 years and 0.8 GtC over 50 years yield 40 GtC, around two-thirds of the "lost" 55 GtC.

The calculations shown in Figure 6-4 assume that between 2000 and 2100 agricultural soils sequester carbon at global rates between 0.4 and 0.8 GtC per annum (see footnote 10). Rates are assumed to be twice as great in the initial years and half as great in the later years. Another assumption is that the full potential of soil carbon sequestration is realized without any additional net cost to the economy. This is not an unreasonable assumption since organic matter added to the soil improves its fertility and productivity. Were the soil carbon sequestration potential of forests and rangelands included in this analysis, longer deferrals would be shown.

3.2. Experimental evidence of soil carbon sequestration

Are the assumed rates of soil carbon sequestration reasonable? Apparently so, since experimental data from the Great Plains locations prove that with appropriate management carbon can be restored to soils of this region. Long-term rotation experiments conducted at Lethbridge and Breton, Alberta show the effects of management on content of soil organic carbon (SOC) at these locations. Figure 6-5a (from Izaurralde et al. 2001) shows an initial decrease in SOC at Lethbridge after cultivation under three rotations—continuous wheat, wheat-wheat-fallow, and wheat-fallow—followed by recovery and stabilization. Figure 6-5b is for a wheat-fallow rotation at Breton with low initial SOC. In these plots SOC either decreases, remains stable or increases in response to crop productivity and nutrient additions. Figure 6-5c representing a cereal–forage rotation (wheat-oats-barley-hay-hay) at Breton, shows increased SOC, particularly where manure was applied. These figures illustrate the effects of rotation, tillage, and nutrient additions on SOC accumulation. Improvement is greatest in the Breton plots where a hay crop is part of the rotation.

Conversion of land from cultivated crops to grasses is particularly beneficial in terms of carbon sequestration in soils. Izaurralde et al. (2006) cite experiments in the US portion of the Great Plains from northern Nebraska to the Texas Panhandle in which, 6 years after conversion from cultivation to grass cover, soil carbon content to the depth of 3 m increased by 2.4–12.5 MgC ha^{-1}. Carbon has also been accumulating in lands planted to perennial grasses under the Conservation Reserve Program (CRP)[11] at rates exceeding 1 Mg ha^{-1} $year^{-1}$ (Gebhart

[11] The Conservation Reserve Program (CRP) is a voluntary program for agricultural landowners in the USA. Through CRP, landowners receive annual rental payments and cost-share assistance to establish long-term, resource conserving covers on eligible farmland. The Commodity Credit Corporation (CCC) makes annual rental payments based on the agriculture rental value of the land, and it provides cost-share assistance for up to 50% of the participant's costs in establishing approved conservation practices. Participants enroll in CRP contracts for 10–15 years. The program is administered by the CCC through the Farm Service Agency (FSA), and program support is provided by Natural Resources Conservation Service, Cooperative State Research and Education Extension Service, state forestry agencies, and local Soil and Water Conservation Districts. (http://www.fsa.usda.gov/dafp/cepd/crp.htm)

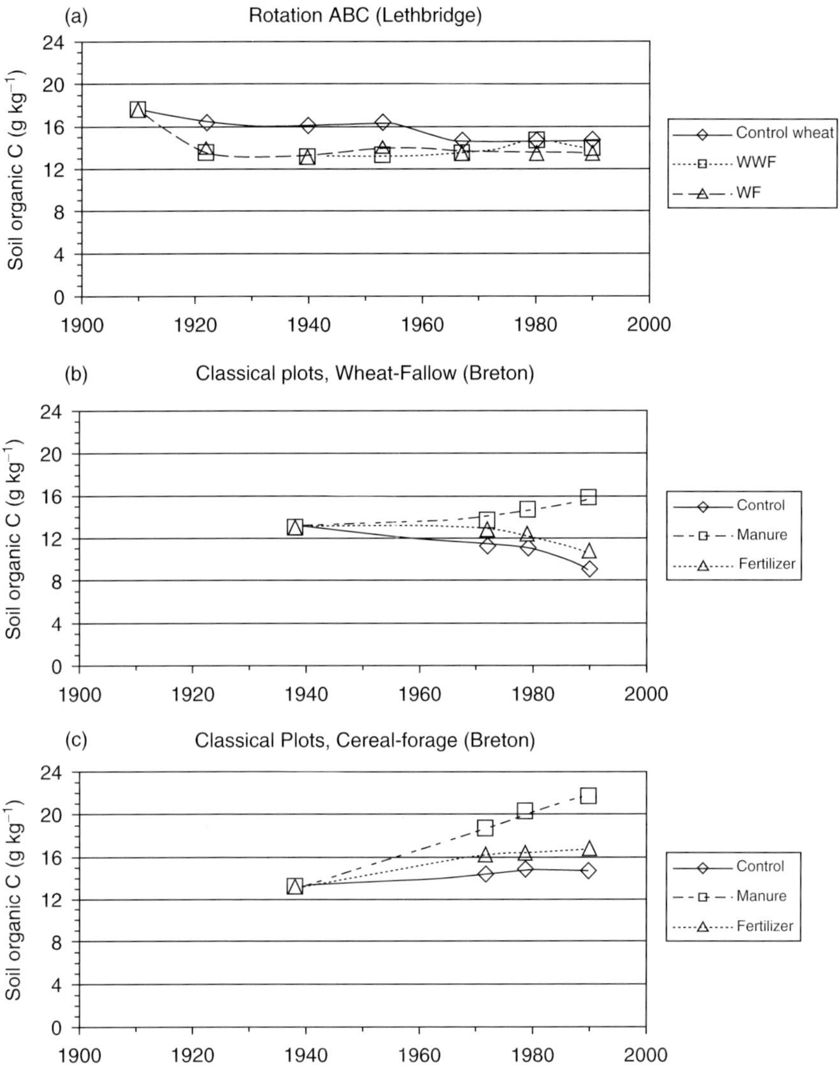

Figure 6-5. Soil organic carbon in two long-term crop rotations in Alberta, Canada (Source: Izaurralde et al. 2001)

et al. 1994). Liebig et al. (2005) report experiments on 42 paired switchgrass (*Panicum Virgatum*) and cropland sites in Minnesota in the northern Cornbelt and in the Plains states of North Dakota and South Dakota in which increases in soil carbon content of 7.74 and 4.35 Mg ha^{-1} were found for the 0.3–0.6 and 0.6–0.9 m depths, respectively. Greater root biomass below 0.3 m in switchgrass was likely the reason for the differences in SOC between the switchgrass stands and cultivated cropland.

In tropical climes even more dramatic sequestration rates may be possible. Fisher et al. (1994) working in the eastern plains of Columbia found that deep-rooted African perennial grasses *Andropogon Gayanus* and *Brachiaria humidicola*, alone and in combination with the legumes *Arachis pintoi* and *Stylosanthes capitata*, native to South America, increased soil carbon content over 6 years from 26–70 Mg ha^{-1} $year^{-1}$. Extended to the full area of similar pastures in South America, these grasses could sequester from 100 to 507 Mt $year^{-1}$.

There is also a substantial literature showing that minimum tillage and no-till management of farm fields in the US Midwest, eastern Great Plains, and Canada leads to increases in soil carbon content (see review by Izaurralde et al. 2001). West and Marland (2002) estimated, on the basis of long-term experiments in the USA a rate of 0.34 MgC ha^{-1} $year^{-1}$ under no-till management in the USA.

However, not all studies show positive changes in SOC with initiation of conservation tillage cropping systems. Verma et al. (2005), for example, found that 4 years after conversion of land from conventional tillage to no-till soil carbon content in 200 kg samples of soil was actually lower by 256, 235, and 116 g C m^{-2} in continuous corn, irrigated maize–soybean, and rainfed maize–soybean rotations near Mead in eastern Nebraska. These differences were not statistically significant and may relate to the preparatory disking of the soil before these no-till rotations were established. Nonetheless these data emphasize the need for caution in projecting major positive mitigation impacts of soil carbon sequestration.

While the picture is not absolutely clear on this matter, a summary of the literature on management effects on soil carbon sequestration by Post et al. (2004) presents convincing evidence overall that intensification of cropping, organic amendments, conservation tillage, establishment of perennial vegetation, and biomass cropping can all contribute to increasing the storage of carbon in soils.

3.3. Soil carbon dynamics

Not all of the carbon in aboveground plant residues or residual roots incorporated in the soil is actually sequestered. Through complex chemical, physical, and biological processes largely involving soil microorganisms, fresh plant materials are either quickly decomposed or converted to short-lived or long-lived forms of humus (organic matter). The microorganism-modulated processes are controlled by temperature and soil moisture conditions. Organic matter is naturally low in well aerated soils of the dry, hot regions and is abundant in colder, wetter regions where organic matter tends to concentrate. In wet soils and swamps the materials are isolated from oxygen and aerobic respiration is quelled. Because the processes of SOM accumulation and decomposition are complex, mathematical models simulating these processes have become useful for predicting how field management (plant type, fertilization, irrigation, tillage, and other processes) will affect organic matter content and also how climate change might affect the processes and resultant sequestration.

A description of soil carbon models is given in Izaurralde et al. (2001). Generally, these models treat carbon flow in soils by dividing SOC into compartments or pools and treating the dynamics as a first order process:

$$dC/dt = -kC + A$$

where dC/dt is the rate of change of carbon concentration (or mass), k is the decomposition rate constant, C is the concentration (or mass) of carbon present at time t and A is the rate of carbon addition (e.g., litter, crop residues).

The number of carbon "pools" or "compartments" may differ among the many published models of soil carbon dynamics. Prominent among these models are CENTURY (Parton et al. 1988) with four compartments, RothC (Coleman and Jenkinson 1966) with two compartments, and *ecosys* (Grant et al. 1993) with eight compartments. The pools in these models can be characterized as active or labile, slowly oxidized, very slowly oxidized, and passive.

Models of soil carbon dynamics are becoming more and more sophisticated and able to address many of the complex management and climatic factors that determine how much carbon can be inserted into the soil and, perhaps more importantly, how much of it will remain sequestered. A good example of such a model is that of Izaurralde et al. (2006) who have adapted the widely used crop growth and yield model Environmental Protection Information Center (EPIC) (described in Chapter 5) to model soil carbon dynamics. These authors use concepts developed in the CENTURY model to connect the simulation of soil carbon dynamics to crop management, tillage methods, and erosion processes. Carbon and nitrogen routines are added that interact directly with soil moisture, temperature, erosion, tillage, soil density, leaching, and translocation functions.

3.4. Model simulations

Based on, and calibrated to, field experiments and an understanding of soil C dynamics, simulation models provide the opportunity to project forward how the effects of management practices, cultivars, new crops, and climate change may, alone or through interactions among them, affect soil carbon content. Figure 6-6 from Izaurralde et al. (2001) shows the results of one such simulation. One hundred years of conventional cultivation of a soil of relatively low initial SOC reduces that quantity from 30 to 26 Mg ha^{-1}. The application of alternative tillage (no-till) initiates a recovery process that, after another 100 years, brings SOC to a level even greater than the initial content.

The interacting effects of tillage, surface residue effects and climate change on SOC storage were simulated by Grant et al. (1998, cited by Izaurralde et al. 2001) using the model *ecosys*, zero tillage increased SOC from ~34 to 35.5 Mg ha^{-1} under

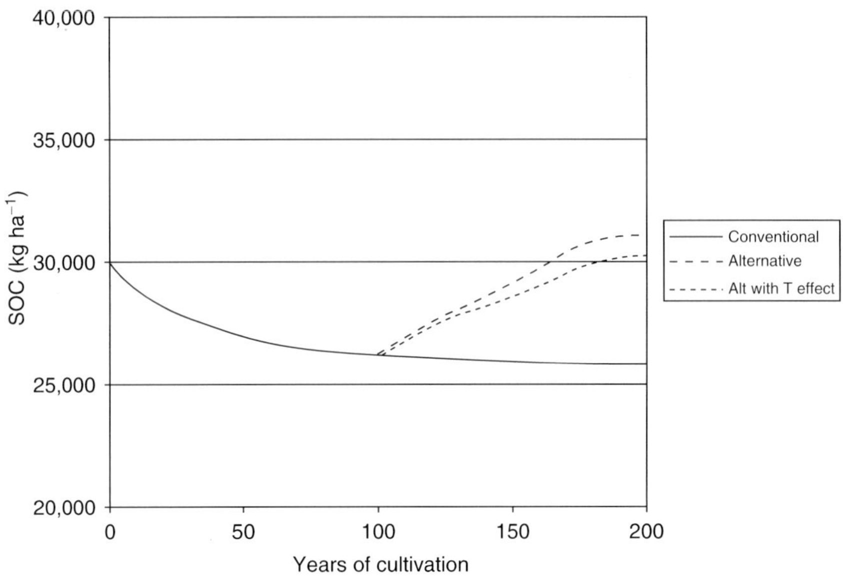

Figure 6-6. Simulated carbon dynamics in a soil with low initial content of soil organic carbon (SOC) with C sequestering practice (Alternative) applied after 100 years of cultivation, with and without the influence of temperature (T effect) on the decomposition rate constant of soil organic matter (Izaurralde et al. 2001)

current climate. Increasing temperature (3°C and 6°C temperature rise) reduced storage considerably more under conventional tillage (~1 and 2.1 Mg ha^{-1}) than under zero tillage (~0.3 and 0.8 Mg ha^{-1}). This simulation assumed that the crops grow over a period of 11 years at a time by which atmospheric CO_2 concentration has doubled from its preindustrial level of 280–560 ppmv. Because the kinetic rate constants associated with the various chemical processes of organic matter (humus) formation and decomposition increase with the increasing temperature, the potential gains in SOC due to no-till are reduced by warming.

3.5. The potential for soil carbon sequestration on the NAGP

In the original process of conversion of the Great Plains from grassland to agriculture, vast quantities of carbon were released from its soils. Further, the forms of tillage practiced in farming these soils led to additional losses of SOM. With proper applications of plant residues, manures and mineral fertilizers SOM can be restored to a modest degree. And with conservation tillage (minimum- and no-till) a substantial restoration of SOM, sometimes to near or even above the original content, is possible. Overall, however, how much additional C can actually be sequestered in the soils of the NAGP? And what technological improvements are needed to permit achievement of maximal sequestration?

Before addressing the capacity of the NAGP, it is instructive to examine some global scale estimates for various land uses. For example, Metting et al. (1999, based on work by Jacobs et al. 1999) estimated potential annual carbon sequestration of 5.65 – 8.71 GtC for all terrestrial ecosystems. Table 6-1 shows their estimate for sustained global terrestrial carbon sequestration potential in ecosystems that dominate the NAGP to range from 3.05 to 3.40 GtC $year^{-1}$.

To accomplish the indicated level of sequestration on agricultural lands will require a high (H) level of best-available management (no-till), intensified production and residue inputs, intensified crop rotations, double cropping, greater use of perennials and new technologies such as precision farming. The calculation for biomass crop lands assumes a high-level management leading to annual aboveground productivity of 13.2 MgC ha^{-1} $year^{-1}$. Belowground carbon storage is 1.75 MgC ha^{-1} $year^{-1}$, assumed to be permanent.

For the grasslands, intensification of management to a medium (M) level involves fertilization, controlled grazing, and species improvements. A 25% increase in belowground carbons stocks results from linear increases through 2050.

On rangelands a total increase of 27 GtC is projected through 2050. This involves rehabilitation of degraded rangeland at a medium level of management and the assumption of an operative CO_2-fertilization effect.

Another comprehensive modeling exercise, that of Jain et al. (2005), provides what should be a realistic assessment for North America as a whole of the two factors—soil management and anticipated climatic change—that might affect its soil carbon balance. This study evaluates how changes in soil management can increase the accumulation of SOC (as a means of mitigating greenhouse warming and climate change) and the extent to which changes in temperature and precipitation and the CO_2-fertilization effect can lessen or augment the amount

Table 6-1. Sustained global carbon sequestration (CS) potential in ecosystems represented on the North American Great Plains. (Abstracted by the author from Metting et al. 1999)[a]

Ecosystem	Primary method to increase CS	Potential CS (GtC/year)
Agricultural lands	Management (H)	0.85–0.90
Biomass crop lands	Manipulation (H)	0.5–0.8
Grasslands	Management (M)	0.5
Rangelands	Management (M)	1.2
Total		3.05–3.40

[a]Based on work by G.K. Jacobs and colleagues at the Oak Ridge National Laboratory and reported in US Department of Energy, 1999. Carbon Sequestration, State of the Science: A working paper for roadmapping future carbon sequestration R&D. US Department of Energy, Office of Science and Office of Fossil Energy. Washington, DC, http://www.fe.doe.gov/ programs/sequestration/ publications/ 1999_rdreport/index.html.

of carbon sequestered. No-till farming is used in this simulation as the surrogate for conservation tillage.

The Great Plains falls primarily in two climatic regions in the classification used by Jain et al. These are cold temperate dry (CTD) and warm temperate dry (WTD). Only the southeastern corner of Nebraska, the eastern quarters of Kansas and Oklahoma, and the northeastern corner of Texas are in a region classified as temperate wet (TW).

As estimated by Jain et al., as of 1765 the initial carbon content of the top 0.3 m of soil ranged from <20 Mg ha^{-1} in parts of Texas, Oklahoma, Kansas, and Nebraska to as much as 80 Mg ha^{-1} or more in portions of the Prairie Provinces. Their modeling of carbon management response suggests that between 1980 and 2000 ~8% of the land may have been converted from conventional tillage to no-till in the CTD region of North America and ~20% in the WTD region. By 2000 no-till had been adopted on ~18% of US cropland as a whole and 30% of Canadian cropland. Jain et al. applied their model using 1980 climate and CO_2 concentrations to estimate how the total of land in no-till affects SOC content and how, if all cropland in North America were converted to no-till that quantity would change.

Climate change increases the rate of carbon sequestration in the CTD lands of North America as a whole from 0.675 to 0.694 MgC ha^{-1} $year^{-1}$. In the WTD region it is increased from 0.570 to 0.595 MgC ha^{-1} $year^{-1}$. With the areas under no-till management in 2000, climate change increases total carbon sequestration from 14.40 to 14.80 TgC $year^{-1}$ in the CTD region and from 4.74 to 4.94 TgC $year^{-1}$ in the WTD region. Were all cropland in North America converted to no-till, sequestration of carbon would increase under climate change from 53.85 to 55.38 TgC $year^{-1}$ and from 17.71 to 18.50 TgC $year^{-1}$ in the CTD and WTD regions, respectively. Overall for North America, the Jain et al. model suggests that no-till alone would have increased soil carbon in the USA and Canada over the period 1980–2000 by 826 TgC and that with climate change and CO_2-fertilization sequestration would increase by an additional 5% to 868 TgC.

Table 6-2 shows results of the application of the Jain et al. model to the Great Plains portion of North America as geographically defined in this work.[12] About 25% of the Great Plains area (USA and Canada) was in some form of no-till agriculture in 2000. Eighty-eight percent of the land area is in the CTD zone and WTD zone. Climate and CO_2 change increase carbon sequestration from 10.93 to 11.25 TgC $year^{-1}$ in the CTD soils and from 3.57 to 3.74 TgC $year^{-1}$ in the WTD soils of the region. As is the case for the continent as a whole, sequestration rates would increase in the NAGP were that region converted entirely to

[12] I am indebted to Professor A.K. Jain and his colleagues at the University of Illinois, Urbana, Illinois for this special run of their model.

Table 6-2. Total cropland area, cropland in no-till (NT), and soil carbon sequestration rates for NT with and without changes in climate and CO_2.[a] (Calculated by the methods of Jain et al. 2005)

Climate regions[b]	Cropland area (Mha)	NT area (Mha)	Modeled experiments[c] (MgC ha^{-1} year^{-1})		Sequestration cases[d] (TgC year^{-1})			
			NTWC	NTWOC	BWC	BWOC	MWC	MWOC
CTD	65.87	17.30	0.65	0.63	11.25	10.93	42.85	41.61
CTM	2.70	0.71	0.17	0.17	0.12	0.12	0.47	0.45
WTD	26.77	7.03	0.53	0.51	3.74	3.57	14.22	13.60
WTM	6.73	1.77	0.23	0.22	0.41	0.40	1.55	1.50
TROP	2.77	0.73	1.03	0.96	0.74	0.70	2.83	2.65
Total	104.84	27.53	2.61	2.49	16.26	15.71	61.93	59.83

[a]The total cropland and NT area is given for the year 2000, while sequestration rates are averaged for the period 1981–2000.
[b]CTD: cold temperate dry; CTM: cold temperate moist; WTD: warm temperate dry; WTM: warm temperate moist; TROP: cold and warm tropics.
[c]NTWC and NTWOC: No-till with and without changes in climate and CO_2.
[d]BWC: Base case with varying climate and CO_2; Base case with constant climate and CO_2 at their 1980 levels; MWC: maximum NT case with varying climate and CO_2; MWOC: maximum NT case with constant climate and CO_2 at their 1980 levels.

no-till farming. In the case of the CTD portion of the Plains soil sequestration would increase from 41.61 to 42.85 TgC year^{-1} and in the WTD portion from 13.60 to 14.22 TgC year^{-1}.

3.6. Soil carbon: a summary

Very substantial quantities of carbon were released from its soils in the original process of conversion of the Great Plains from grasslands to farming. Further, the forms of tillage practiced in farming and grazing these soils led to additional losses of SOM and its carbon. Research at many sites on the Plains has shown that SOM can be restored to a modest degree with proper applications of plant residues, manures, and mineral fertilizers. With conservation tillage (minimum- and no-till) and, especially, with conversion from cultivated crops to grass cover, a substantial restoration of SOM is possible. Simulation models incorporating knowledge of soil carbon dynamics track observations and permit projections of how future management can favor continued restoration of SOM and carbon, sometimes to near or even above the original content. Further, the simulation models allow reasoned projections of how future climatic conditions and atmospheric CO_2 concentrations may limit or enhance carbon sequestration in soils. The Jain et al. (2005) model projections suggest that these conditions could increase sequestration slightly on the Plains. Overall, it appears that soil carbon sequestration can improve prospects for sustainable agriculture on the Plains and can contribute modestly, especially in the near term, to mitigation of greenhouse warming.

4. BIOMASS CROPS

4.1. Background

Biomass, plant matter of recent (nongeologic) origin, has been used for fuel and other purposes from time immemorial. Traditional biomass as typically used today in developing countries includes wood and herbaceous materials gathered and brought to villages for cooking and heating purposes. Manure, too, remains a common source of fuel throughout the less developed world. Indeed, on the Great Plains buffalo and cow "chips" were used as fuel by both the Native Americans and early European-origin settlers. With low grain prices during the depression of the 1930s farmers on the Plains, making use of another form of biomass, burned their corn for fuel.

There are many modern forms of biomass in commercial use today, most stemming from agricultural and forest wastes (Lynd et al. 2003). The primary sources according to a classification by Oak Ridge National Laboratory (ORNL) (Perlack et al. 2005) include crop residues from land managed for grain, forest residues from land managed for timber and pulp and from land thinned for fire protection. Secondary sources include manure and biosolids,[13] mill wastes, and black liquor, a pulp by-product. Another consequential source of biomass is urban wood waste from construction and demolition sites.

The Biomass Research and Development Act of 2000 (US Congress 2000) created a Biomass Research and Development Technical Advisory Committee to advise the Secretaries of Agriculture and Energy on program priorities and to facilitate cooperation among various federal and state agencies and private interests. This group set a challenging national goal for biomass to supply 5% of the nation's power, 20% of its transportation fuels, and 25% of its chemicals by 2030—a fivefold increase over consumption reported by the DOE in 2003. How these goals might be realized is discussed below.

4.2. Current and future contributions of biomass to US energy supply

In 2003 nonrenewable energy sources provided 94% of US requirements: petroleum contributed 39%; natural gas, 24%; coal, 23%; and nuclear, 8%. Only 6% was contributed by renewable sources and of that 6% biomass and hydroelectric facilities contributed 47% and 45%, respectively. Geothermal, wind and solar together contributed only 8%. Thus, biomass contributed only 3% of the then current US energy consumption.

Actual annual consumption of biomass feedstock for bioenergy and bioproducts together was ~172 million dry tonnes (Mt_d) (Perlack et al. 2005). The primary sources of biomass in the USA are forests and agriculture, the latter with

[13] Solid, semisolid, or liquid residues generated during primary, secondary, or advanced treatment of domestic sanitary sewage. (http://www.michigan.gov/deq/0,1607,7-135-3313_3683_3720-9573--,00.html)

the potential almost three times greater than the former (905 and 334 Mt_d per annum) as shown in Tables 6-3 and 6-4 for forests and agricultural sources of biomass, respectively.

The assumptions underlying Table 6-3 are conservative: (1) that all forested areas not currently accessible by roads are excluded; (2) that all environmentally sensitive areas are excluded; (3) that equipment recovery limitations are considered; and (4) that recoverable biomass is allocated to two utilization groups—conventional forest products and biomass for bioenergy and bio-based products.

The estimates of biomass from agricultural sources in Table 6-4 are based on certain key optimistic (but also realistic) assumptions. These are: (1) that yields of corn, wheat, and other small grains increase by 50%; (2) that the residue-to-grain ratio for soybeans increases from 1.5–1 to 2.0–1; (3) that harvest technology becomes capable of recovering 75% of annual crop residues (when removal is sustainable); (4) that all crop land is managed with no-till methods; (5) that 22 million hectares of cropland, idle cropland, and cropland pasture is dedicated to the production of perennial bioenergy crops; (6) that all manure in excess of that which can be applied on farm for soil improvement under anticipated EPA restrictions is used for biofuel; and (7) that all other available residues are utilized.

The aforementioned DOE/US Department of Agriculture (USDA) report of Perlack et al. (2005) asserts that "...The biomass resource potential identified can be produced with relatively modest changes in land use and agricultural and forestry practice" and, further, that this "potential" may not be an upper limit but is, rather, one scenario based on a set of reasonable assumptions. Future advances in science and technology, discussed below, could well increase this

Table 6-3. Potential sources of forest biomass (Perlack et al. 2005)

Source	Millions of dry tonnes per annum
Fuel wood harvested from forests	47
Residues from wood processing mills and pulp and paper mills	132
Urban wood residues including construction and demolition debris	43
Logging and site clearing operations	58
Fuel treatment operations to reduce fire hazards	54
Total	334

Table 6-4. Potential sources of agricultural biomass (Perlack et al. 2005)

Source	Millions of dry tonnes per annum
Annual crop residues	388
Perennial crops	342
Grains used for biofuels	79
Animal manures, process residues, and other miscellaneous feedstock	96
Total	905

Table 6-5. Current and anticipated biomass contribution to national sector needs (Source: Perlack et al. 2005)

Sector	Units	2001	2010	2020	2030
Biopower	%/(Exojoules)	–/(2.84)	4/(3.38)	5/(4.119)	–/(5.06)
Transportation fuels	%/(Exojoules)	0.5/(0.02)	4/(1.37)	10/(4.22)	20/–
Biobased chemicals and other products	%/Mt	5.0/(5.67)	12/–	18/–	25/–

potential substantially. In this chapter we explore the question of how much biomass can be produced sustainably on the Great Plains. Table 6-5 shows the Perlack et al. time course for achieving the goal of sustainable biomass production.

4.3. The Great Plains contribution

4.3.1. Forests as a source of biomass

The distribution of land use by area in the Great Plains states was shown in Table 2-4. Over 41 million hectares are forested in one way or another. The bulk of the forested land in these states is found not on the Plains themselves but rather in the Rocky Mountain portions of Montana, Wyoming, Colorado, and New Mexico. The eastern portion of Oklahoma (outside the boundaries of the Plains as we have defined them) accounts for most of its forests.

Hardwoods grow in the eastern portions of Nebraska, Kansas, South Dakota, and North Dakota. Prominent species are the poplar and cottonwood (*Populus deltoids*), blackjack oak (*Quercus marilandica*), and Osage orange (*Maclura pomifera*). In the western portions of these states softwoods—predominately pine—appear.

Reforestation (restoration of forest vegetation to lands previously deforested) and afforestation (plantation of trees where, at least in historic times, they had not grown before) have been proposed as a means of reducing CO_2 accumulation in the atmosphere.

Afforestation is not a new concept for the plainsman and woman. In presettlement times the Plains were, of course, dominated by tall and short grasses with trees growing mostly in and along water courses. Almost immediately on their arrival, settlers from the forested eastern regions of the continent began to plant windbreaks to protect their farmsteads and shelterbelts to protect their crops. Now known as the "Cornhusker State," Nebraska was earlier described as the "Tree Planter State."

The history of governmental efforts at afforestation in the Plains is well described by Hurt.[14] These efforts involved the planting of "national forests,"

[14] "Forestry on the Great Plains, 1902-1942." Lecture presented in 1995 at Kansas State University by Professor R. Douglas Hurt, History Department, Purdue University. (http://www-personal.ksu.edu/~jsherow/hurt2.htm)

seedling nurseries and shelterbelts in the drier regions of Nebraska and Kansas. President Franklin D. Roosevelt took office in March 1932. He was a strong believer in the environmental merits of tree planting and, early in his first term, recognized that a massive tree planting effort on the Plains would provide much needed work opportunities for the unemployed in that depression era.

Planning for a "shelterbelt project" began in 1933. Roosevelt favored planting a 100-mile wide forest from the Canadian to the Mexican borders. Government foresters favored planting shelterbelts at 1-mile intervals in a 100-mile wide zone. Many variations were made in the latter scheme and the complete grid envisioned was never completed. Nonetheless the work of the "Prairie States Forestry Project" actually began in 1934 and during the ensuing 8 years over 30,000 shelterbelts were planted to reduce wind erosion, capture snow, reduce evaporation, and to protect crops and animals from the strong drying winds of the region. These shelterbelts were comprised of some 220 million trees and spread across 18,600 miles (28,800 km) of open fields on 30,000 farms. Work on the shelterbelt project was interrupted when the USA entered World War II.

By now many of the 1930s-era shelterbelts have been thinned or removed entirely to facilitate irrigation and the use of farm machinery very much larger than anything in use when the trees were planted. Yet the planting of shelterbelts continues at this time with guidance based on research conducted by the USDA Forest Service and Land Grant universities and with the assistance of state forestry agencies. In Canada, the Prairie Farm Rehabilitation Administration continues to operate similar programs.

In spite of these monumental efforts to afforest the Plains, in terms of land use the region still remains largely treeless. In the US portion of the Plains proper, an area more than 202 million hectares in extent, less than 4 million hectares is forested, whether in public, industrial, or private ownership.

Thus it would appear that forest land can contribute only little—at any rate far less than the current agricultural and grazing lands—to biomass production in the region. But with selective harvest, brush clearance for fire avoidance, and thinning management of windbreaks and shelterbelts, these lands might provide a small but constant supply of biomass.

The potential role of trees for sequestering carbon and as a biomass crop is explored in subsequent sections of this chapter.

4.3.2. Agricultural sources of biomass

Agricultural sources of biomass are of two kinds: (1) grains and residues from annual crops and (2) dedicated perennial crops. Grains and oilseeds are the primary feed stocks used to produce ethanol, biodiesel and by-products today (Perlack et al. 2005). Together with processing residues from foods and feeds and manure and other residues these materials account for ~25% of current biomass consumption. The potential for agricultural sources of biomass is considered to be much greater than this. The total current sustainable availability of biomass from cropland in the USA is ~176 Mt_d per year (Table 6-6,

Table 6-6. Potential sources of sustainable biomass from agricultural lands in the USA (millions of dry tonnes per annum)

	Current	With technological change	
Source	(ca. 2001)	Moderate	High
Corn stover	68.0	154.2	240.4
Wheat straw	10.0	31.8	51.7
Small grain residues	5.4	13.6	22.7
Other crop residues	19.1	33.6	43.5
Grains to biofuels	13.6	50.8	88.0
Manures	31.8	40.0	39.9
Other residues	28.1	36.3	39.9
Subtotal	176.0	360.2	526.1
CRP biomass		25.4	25.4
Total		385.6	551.5

based on Perlack et al. 2005, Figures 17 and 18) of which slightly more than a fifth is currently used.

The US potential sustainable agricultural sources of biomass increase from 176 to 360 and 518 Mt_d per year with estimated effects, respectively, of moderate and high crop yield increases over the coming decades. This estimate assumes no change in land use. These numbers do not include another 25 Mt_d of biomass harvested on land under the CRP.

The increases in potential biomass are predicated on improved technologies—crop yields for corn increased by 25–50% and for wheat and other small grains, soybeans, rice, and cotton at lower, yet significant, rates. Without specific plant breeding for that purpose yield increases may not alter the harvest index (ratio of the harvested portion to the total plant above ground biomass) to any great extent. Collection equipment is assumed capable of recovering as much as 60–75% of the crop residues under moderate and high yield increase, respectively. It is further assumed that no-till cultivation is practiced on ~81 million hectares of cropland under the moderate yield increase scenario and on all active cropland under the high yield scenario. Other assumptions supporting this analysis are: (1) that food, feed, and export requirements are met before determining the amounts of corn and soybeans available for ethanol, biodiesel, or other bioproducts, providing 3 and 5 times more than 2001 levels under the moderate and high-yield scenarios, respectively and (2) that ~68 Mt_d of manure and other agricultural wastes and 50% of the biomass produced on CRP lands are available for bioenergy production.

Perlack et al. (2005) consider these goals eminently "do-able." It is clear, however, that commitment to research and development, deployment of new technologies and incentives to farmers and the energy-consuming sectors will be required if these potentials are to be reached.

4.4. Biomass crops for the Plains

What are the most likely perennial biomass crops for the Great Plains? A list of herbaceous and woody species suitable for biomass production and the criteria for their selection are given by Tuskan et al. (2004). Among the grasses that have been studied thus far are: switchgrass (*P. virgatum*), big bluestem (*Andropogon* spp.), tall fescue (*Festuca* spp.), and reed canary grass (*Phalaris* spp.). Another potential bioenergy crop is miscanthus (*Miscanthus* × *giganteus*) which has been grown from rhizomes successfully in a number of European countries (Lewandowski et al. 2003) and has also been tested in Illinois.[15] This species may also prove useful in the wetter, warmer portion of the Plains region although experience with it on the Great Plains is lacking, except as an ornamental. Among the woody species that can be grown in short rotation are the cottonwood (*P. deltoides*) and other members of that species, the sycamore (*Plantanus* spp.) willow (*Salix* spp.) and others such as sweet gum (*Liguidandar* spp.) and black locust (*Robinia,* spp.).

Lynd et al. (2003) consider that for biomass purposes the grasses have some distinct advantages over the woody crops. The grasses are of similar or higher productivity than the woody crops; they are compatible with a broader range of sites, especially dryer sites; farmers know how to grow grass and have the equipment to do so; since the grasses are harvested at least once each year, production is generally more easily financed; grass production is generally accompanied by carbon sequestration; and grasses are more readily incorporated into crop rotations.

4.4.1. More on the woody species

Short rotation woody crops (SRWC) hold the promise of providing reliable supplies of biomass for use directly as biofuel and indirectly for conversion into ethanol and other bioproducts. The genus *Populus* includes several species native to North America. Tuskan et al. (2004) describe the *Populus* species in North America as

> dioecious (having male and female flowers on separate plants of the same species), deciduous pioneering species established either through seed or vegetative propagation on bottomland sites or, in the case of *P. Tremuloides*, on burned areas. *Populus* species are fast growing with relatively short life spans, reaching maturity at about 30+ years. Commercial plantations use inter-specific hybrids in conjunction with SRWC silvacultural practices.

[15] http://www.eurekalert.org/pub_releases/2005-09/uoia-hgm092705.php. Miscanthus, sometimes confused with elephant grass (*Pennisetum purpureum*), a tall perennial grass, has been tested in Europe during the past 5–10 years as a new bioenergy crop. Miscanthus can be harvested every year with a sugar cane harvester and can be grown in a cool climate like that of northern Europe. Like other bioenergy crops, the harvested stems of miscanthus may be used as fuel for production of heat and electric power, or for conversion to other useful products such as ethanol. (http://bioenergy.ornl.gov/papers/miscanthus/miscanthus.html)

According to Tuskan et al. (2004) these silvacultural practices include the use of genetically improved, clonally propagated plants established and managed on agricultural quality land. Site preparation may be extensive and weed control, pest management and fertilization is practiced as needed. For biomass or pulping, rotations are typically 5–10 years in duration. On nonirrigated lands growth rates vary from 6 to 21 Mg ha^{-1} year^{-1} and up to 30 Mg ha^{-1} year^{-1} on irrigated plantings.

For biomass production the optimum poplar tree under short-rotation silviculture would put most of its carbon into the trunk, would have a compact crown and root system and a high harvest efficiency. If "genetically engineered," the tree would be nonflowering so as to eliminate the possibility of gene flow out of the plantation. Additionally, a nonflowering clone would increase allocation of biomass to the stem.

Despite its potential importance in other regions, the general consensus of specialists in woody biomass[16] is that without irrigation the poplar, willow and other species mentioned above would not be competitive with the grasses as a sustainable source of biomass on the Great Plains.

4.4.2. Switchgrass

The USDOE selected switchgrass as a model for its research on herbaceous crop species for bioenergy. Switchgrass is a warm season perennial whose native range extends from Central America to much of eastern Mexico and Baja California, over most of the eastern two-thirds of the USA, including Arizona, New Mexico, and the Great Plains portions of Colorado, Wyoming, and eastern Montana and into Manitoba, Ontario, New Brunswick, and Quebec.[17]

Switchgrass was an important component of the native, highly productive North American tallgrass prairie. Now native stands of switchgrass in the USA and Canada are few and widely scattered as that grass, along with other natives, was plowed up as the Plains and Prairie Provinces were converted to agricultural uses.

A review by McLaughlin et al. (1999) describes the characteristics of switchgrass and research efforts being made to establish its utility as a biomass crop:

> It is high yielding, is useful in soil conservation and is compatible with conventional farming practices. Switchgrass tolerates diverse conditions from arid sites in the shortgrass prairie to brackish marshes and open woods. Switchgrass is open-pollinated, has a very deep well-developed rooting system and has the photosynthetically efficient C4 metabolism. It reproduces both by seeds and vegetatively. Biomass in

[16] Personal communication, Dr. S.D. Wullschleger, Oak Ridge National Laboratory, April 7, 2006.

[17] http://www.EAP.MCGill.CA/MagRack/SF/Fall%2091%20L.htm. With ECU support switchgrass is also being studied in Europe for biomass production. (http://www.nf-2000.ORG/secure/Fair/S817.htm)

> the root system may exceed that above-ground, giving it an advantage in accessing water and nutrients—a trait that helps it survive under stressful conditions. Switchgrass is ecologically diverse. Two major ecotypes occur: one a thicker stemmed lowland type, adapted to warmer, moist habitats in its southern range; the other a finer stemmed upland type found typically in mid-to northern areas (Vogel et al. 1985). The Dept. of Energy's switchgrass research program began in 1992 and involves universities, DOE sponsored laboratories and field testing sites operated by the U.S. Department of Agriculture's Natural Resources Conservation Service. The U.S. switchgrass research program includes field trials and testing sites, breeding activities, basic research on tissue culture and physiology and genetics.

Switchgrass has been tested in the eastern and central portions of the Plains. As of 2004, switchgrass field trials were being conducted at six locations in Texas, at least two of which fall within the Great Plains region, at one site each in eastern Kansas and Nebraska, five sites in South Dakota, and three in North Dakota. Breeding projects were underway in Oklahoma and Nebraska.[18]

Agronomic research over the last decade and a half has shown that switchgrass can be grown in 10-year rotations. During the first year the grass develops its extensive root system with as much biomass ultimately underground as it supports aboveground. The crop may be harvested late in its second year and once or twice in each subsequent year until it is removed to make way for other crops in the rotation.

Depending on location, current field-scale annual yields of switchgrass range from 9 to 23 Mg ha^{-1}. Yields of six switchgrass cultivars in variety trials at the MacDonald Campus Farm in Montreal, Quebec ranged from 8.9 to 10.9 Mg ha^{-1}.[19] A considerable amount of field testing has been done with switchgrass in the US Southeast, Gulf Coast, and Arkansas. McLaughlin and Kszos (2005) report average yields of adapted varieties over all test sites in this region to range from 10.7 to 23.0 Mg ha^{-1}.

Yields in the Great Plains states of Kansas, Nebraska, and North Dakota also vary widely (9.5–20.6 Mg ha^{-1}) depending on variety. The theoretical yield according to the ORNL's ALMANAC model (Kiniry et al. 1992) is 51.6 Mg ha^{-1}. With successful breeding and agronomic research, reliable field-scale yields of 15–22 Mg ha^{-1} are expected within 20 years. Biotechnology may be the key to breakthroughs that will allow this potential switchgrass yield to be approached (see Section 6.5).

From research by Duffy and Nanhou (2002) on costs of switchgrass production in southern Iowa, we can gather information on the agronomic practices

---

[18] L. Wright. Briefing on Energy Crop Resources and Technologies, National Commission on Energy Policy (NCEP) forum. Future of Biomass Transportation Fuels. June 13, 2003.

[19] P. Girovard, Research update. (http//www.EAP.McGill.CA/MagRack/SF/Summer%2094%20c.htm)

used to establish that crop. Switchgrass can be seeded in fall (frost seeding) or in spring. Land preparation depends on the previous land use. Generally discs, harrows, fertilizer spreaders, and planters used for standard crops can be used to prepare and seed the crop. For no-till management seed can also be drilled directly into the seedbed on land previously under crop production or on land under grass production or pasture. Chemical herbicides can be used in the preparation of both conventional and no-till plantings. USDA/Agricultural Research Service (ARS) and State Extension Service Reports indicate that in field plot research in Nebraska and the Dakotas yields increase year by year for the first 5 years from seeding and stand establishment.

Nitrogen is needed to sustain a switchgrass stand. Vogel et al. (2002) report on tests in Iowa and Nebraska during the 1990s in which nitrogen was applied at the rate of 120 kg ha^{-1} in the first and third years of stand establishment. Optimal biomass yields were obtained when the switchgrass was fertilized at this rate and harvested in a 3-week period after plants were fully headed. Ten to 12 kg N was required to produce 1 Mg of biomass yield.

Despite glowing reports of its productivity and manageability, switchgrass is not yet a "supercrop." It can be as susceptible to disease and insect attack as other grasses, depending on weather and the intensity of pathogen and insect infestation. Gravert et al. (2000) report, for example, on a "smut" (*Tilletia maclaganii*) attack in southern Iowa that reduced switchgrass biomass yields by more that 50%.

Until recently, switchgrass has been domesticated primarily to improve its quality as forage. As an energy crop, however, some genetic modification of its characteristics will be necessary. High density stands of uniform thick tillers[20] with leaves organized to optimize capture of sunlight are required. Flowering, which normally ends the period of biomass accumulation, must be delayed or eliminated to increase allocation to the cell walls. If genetically engineered, suppression of flowering would also eliminate the spread of viable seed or pollen into native populations. More on this matter below.

4.4.3. Grasses for the western Plains

4.4.3.1. Switchgrass out West Because of lower precipitation and its greater interannual variability the western Plains (arbitrarily west of the 100th meridian) is less likely to produce consistently high yields of switchgrass than is the eastern of portion of the region. However, research conducted in the vicinity of the 100th meridian in South Dakota by Lee and Boe (2005) and Mulkey et al. (2006) and in North Dakota by Berdahl et al. (2005) does indicate potential for that crop there.

Lee and Boe (2005) grew two switchgrass cultivars near Pierre in central South Dakota (44°N, 100°W). These were Dacotah, originating at 44°N, 100°W, and Cave-In-Rock, originating to the southeast at 37°N, 88°W). More than 4 years

---

[20] Shoots emerging from the root or bottom of the original stalk.

of study yields for both cultivars varied from more than 9 to less than 2 Mg ha^{-1}. Maximum yields of Dacotah were obtained with July/August cuttings. Cave-in-Rock provided maximum yields as late as September, depending on amount of summer precipitation. It was also noted that biomass yields of overwintered stands harvested near the ground were 85–99% that of stands cut 10 cm aboveground at the end of the previous growing season. This suggests that biomass can be "stockpiled" over winter to trap snow and provide wildlife habitat. It would seem that such a practice would also help regulate the year-round flow of biomass to processing facilities or power plants.

Also in South Dakota, Mulkey et al. (2006) studied switchgrass management in conjunction with CRP requirements at three locations near and east of the 100th meridian. Specific objectives of the research conducted at sites near 44°N, 97°W; 46°N, 97°W, and 44°N, 100°W were to: determine the effects of harvest timing and nitrogen application rates on biomass production and characteristics of switchgrass on land enrolled in the CRP, or with similar management, and to evaluate the impact of harvest management on species composition and persistence. Switchgrass grown at these three sites had not, it appears, been previously harvested.

Harvesting after killing frost produced higher yields than harvesting at the anthesis stage and improved switchgrass stand persistence. Moderate fertilization (56 kg N ha^{-1}) proved optimal. Another benefit of late harvest is that fiber and lignin content of the switchgrass increased between anthesis and killing-frost harvests while total nitrogen and ash decreased.

The Mulkey et al. study demonstrates that switchgrass can be managed on CRP land in ways that are consistent with that program's objectives. Stands on CPR lands may be subjected to every-other-year harvesting (i.e., 50% cut, 50% left for snow catch and wildlife habitat). Taking annual weather-driven variability of production into account, fields may also be harvested in other fractions (i.e., 1/4, 1/3, 2/3 cut) so as to provide a more constant flow to biomass processors. This form of management may provide flexibility to biomass producers and processors.[21] Under such a scheme farmers would continue to receive some benefits from CRP payments while they gain income from the sale of biomass.

Switchgrass was also studied at two sites west of the 100th meridian near Mandan (46°N, 103°W) in North Dakota and at a third site ~265 km further to the west. Yields pooled over the three sites and 3 years ranged from 4.8 to 7.1 Mg ha^{-1} when the crop was cut in August and 5.0–7.8 Mg ha^{-1} when cut in September. Survival percentage ranged from 61–96% to 65–96% for August and September cuttings, respectively. Variations other than those attributable to characteristics and provenance of the cultivars were associated with weather conditions. "Sunburst," a cultivar from southern South Dakota, was the best yielder at all sites and during each year. At the site with the highest soil productivity Sunburst yielded

[21] Personal communication, Professor V.N. Owens of South Dakota State University. April 26, 2006.

12.5 Mg ha^{-1} in a year of above average precipitation and 3.2 Mg ha^{-1} in a drought year. Biomass yields of adapted switchgrass cultivars fluctuated widely at these sites in western North Dakota, depending mostly on available soil water.

4.4.3.2. Other grasses for the western Plains Discussion thus far has been limited to experiments with switchgrass. While switchgrass-for-biomass research has dealt with management strategies for stand establishment, optimal fertilization practices, timing of harvests, etc., other candidate grasses for the western Plains have been studied primarily for their forage potential. In addition, a substantial effort has been devoted to improving yield and biomass quality characteristics in existing cultivars of switchgrass by traditional plant breeding methods and the switchgrass genome is being decoded. But little has yet been done to breed other grasses for the same purposes. Thus there is very little information on yield of grasses other than switchgrass that could be potentially useful were they to be managed for optimal biomass production.

What other grasses have the potential for biomass production in the western and northern portions of the plains? Among those now under study are big bluestem (*Andropogon gerardii*) and Indiangrass (*Sorghastrum notans* (L.) Nash), both of which, like switchgrass, are warm season C4-metabolism species. Others candidates include cool season C3-metabolism species such as intermediate wheatgrass (*Elytrigia intermedia* Nevski spp.), smooth bromegrass (*Bromus invermis*) and crested wheatgrass (*Agropyron desertorum* (Fisch × Link) Schult.). Observations show that the C3 grasses are more easily established than the C4 grasses in the dryer western Plains. Agronomists who have studied these species think that their average yields are likely to be in the 4–6 Mg ha^{-1} range but yields will be very low in years of sparse precipitation.

A consortium of state and federal agencies in North Dakota has recognized the need for a comprehensive evaluation of a wider set of potential biomass crops. At this writing a 10-year study is being initiated to identify appropriate grass species, harvest methods, and other management practices needed to maintain productive perennial biomass stands in the northern and western Plains region.[22] The study is motivated not only by the prospects of biomass as a source of revenue for the farmer but also by the need for a more sustainable agriculture on the more than 2.8 million hectares of highly erodible and saline soils in the state. The study will also provide information on the economics of bioenergy crops and their impacts on organic matter content and carbon sequestration in soils.

The study is to be conducted at five sites in western North Dakota. These sites lie between 46°01′ and 48°14′ N and between 99°07′ and 103°38′ W.

[22] Personal communication, Arnold Kruse, N. Dakota Game and Fish Department, May 5, 2006. Other agencies participating in this project include the North Dakota Natural Resources Trust, the North Dakota State University Extension Service, Research Experiment Stations and the North Dakota Commerce Department.

Three cultivars of switchgrass and one cultivar each of tall wheatgrass (*E. pontica* (Podp.) Holub ssp.), intermediate wheatgrass, big bluestem and Alti wildrye (*Elymus angustus* Trin.) will be grown at these sites alone or in combinations. In addition the wheatgrasses will be grown in combination with alfalfa (*Medicago sativa* L.) and sweet clover (*Melilotus*) for use specifically on CRP land. Annual and biennial harvest regimes will be tested on each of the ten mixtures to be tested.

Can the western Plains produce a reliable supply stream of biomass to support a processing industry? A positive reply to this question will require convincing evidence that the region is able to produce a sustainably high average yield on sufficient acreage within a given radius of the power plant or cellulosic ethanol plant. The "conventional wisdom" (hardly an appropriate term for any wisdom on a topic as new as biomass cropping) has been that switchgrass production on the Great Plains would be limited by supply-stream requirements to the lands east of the 100th meridian in which processing facilities can draw supplies from within a radius of 80 km or so. In view of its semiarid climate, it may perhaps be unrealistic to expect the western Plains to produce enough biomass to satisfy that particular criterion.

However, rapidly increasing world energy prices will, by increasing demand for alternatives to petroleum-based fuels, improve economic prospects for biomass wherever it is produced. It is not unrealistic to assume that the costs of transporting biomass for longer distances in the western Plains may be accommodated as rapidly rising energy costs begin to justify higher prices for cellulosic ethanol and electric power. In addition to rising energy cost, changes in the global distribution of grain production and discontinuation of agricultural subsidies (should that ever become politically feasible) could make the production (especially under irrigation) of crops such as corn and wheat economically impractical on the Plains. The existing irrigation infrastructure might then be used for supplemental irrigation of biomass crops. McLaughlin and Kszos (2005) suggest that "switchgrass yields may be increased substantially in arid areas by low frequency irrigation that could involve equipment rotation among fields." Of course, such an approach, if economically feasible, might also be applied to grasses such as those mentioned above that are more typical than switchgrass of the western Plains.

4.5. Biomass crops under changing climate

The primary interest in switchgrass today is its potential in helping to mitigate climate change. But switchgrass may have an added advantage under Great Plains conditions in that it is likely to be better adapted to anticipated climate changes than are the traditional crops it might replace. Brown et al. (2000) used the EPIC crop growth model to simulate the production of corn (*Zea Mays* L.), sorghum (*Sorghum bicolor* (L.) Moench), soybean (*Glycine max* L.), winter wheat (*Triticum aestivum* L.)—all annual crops—and the perennial switchgrass at 302 sites in the

Missouri–Iowa–Nebraska–Kansas (MINK) region under both current climatic conditions and a Global Circulation Model (GCM)-derived scenario of possible climate change. This scenario was produced using the Commonwealth Scientific and Industrial Research (CSIRO) GCM (Watterson et al. 1995) to drive the National Center for Atmospheric Research (NCAR) regional CM2 model (RegCM, Giorgi et al. 1998) and involved warming of from 3°C to 8°C across the region and increasing precipitation, particularly in the western portion of the region. Crop production was simulated at two atmospheric CO_2 concentrations, (1) 365 and (2) 560 ppmv, the first to represent the absence of, and the second the presence of, a CO_2-fertilization effect (described in Chapter 5).

In general, yields of corn and soybeans declined sharply under the lower CO_2 concentration because of frequent heat stress and early crop maturation. Wheat benefited slightly because of milder winters and less cold stress. The range of simulated switchgrass yields under current (baseline) climate conditions was 2.2–12.0 Mg ha^{-1}. Under the RegCM scenario alone the range rose to 3.8–17.5 Mg ha^{-1} and with CO_2-fertilization and RegCM it rose further to a range of 4.0–19.7 Mg ha^{-1}. The overall increase in switchgrass yields at all sites in response to climate change is a function of the shorter winters, a reduction in the duration of winter dormancy and vegetative growth continuing much later into the fall because of the delayed arrival of killing frosts. Yield losses of the traditional annual crops and switchgrass were offset in this modeling study by the CO_2-fertilization effect at the higher concentration and exceeded those under baseline climate conditions. Switchgrass yields were further increased under CO_2-fertilization.

The precipitation increases associated with the particular climate change scenario used in the Brown et al. study increased soil erosion under the traditional annual crops but not under the dense switchgrass cover, except in the establishment phase, because of its increased growth, longer growing season and the permanent cover it provides. Another environmental benefit was identified: nutrient stress was virtually eliminated on all model farms under switchgrass cultivation as the higher soil temperatures speeded up nitrate formation and increased crop-available N.

5. HOW GENETIC ENGINEERING MIGHT HELP

5.1. Background

Two apparently conflicting imperatives have been identified: (1) the need to divert land to the production of biomass crops as a substitute for some fraction of current fossil fuel use; and (2) the need to maintain and possibly increase supplies of food, feed, and fiber for a growing world population.

Since agriculture began 10,000 or more years ago, farmers have been searching their plots or fields for individual plants that display desirable traits—more or larger seeds, rapid germination, resistance to disease, insect attack, frost, etc.

The seeds of these plants have been grown with care to increase them in number and, little by little, the favored plants replaced the inferior ones. Mendelian genetics provided the basis for a more efficient means of directly introducing desirable traits into the existing populations by cross-breeding. The traditional selection and breeding process is slow and tedious as the first generation progeny of these crossings carry undesirable as well as desirable traits. New cultivars must be backcrossed many times before the undesirable traits are suppressed and only the desirable ones remain. Another approach used since the early 20th century has been the induction of genetic mutations in plants by various physical and chemical means and, since the 1950s, primarily by means of irradiation. Of the many mutations that result from these practices only a very few have the sought-after traits and the plants that do must be crossed with others before the desirable trait can be fixed in a usable cultivar.

Crop yields in the developed world have increased almost continuously since the end of World War II, largely because of active plant breeding programs supported by governments and the private sector to enhance desirable traits in crop and forage plants, although the increased use of chemical fertilizers and pesticides has also been an important factor. A typical example of ever-increasing yields is shown for the US corn crop in Figure 6-7.

One convincing demonstration of the impact of "traditional" plant breeding (in this case hybridization of corn) appears in Figure 6-8 from a report of the Council for Agricultural Science and Technology (CAST 1992) in which hybrids introduced between 1930 and 1984 were grown during a wet year (1987) and a dry year (1985). The figure clearly shows that in both cases yields increase almost linearly with newness of the hybrid.

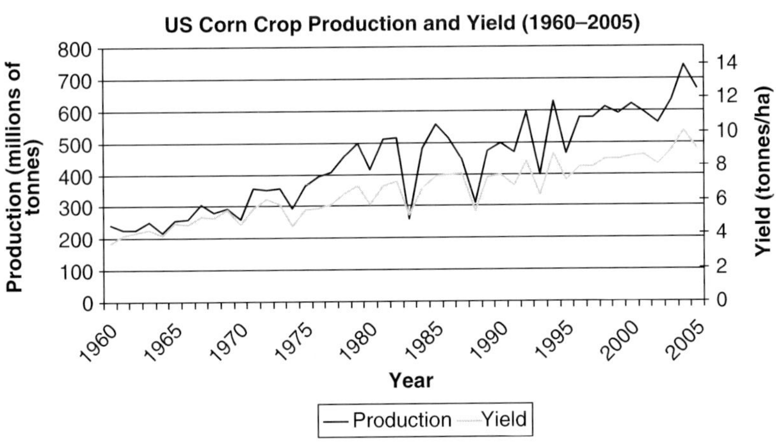

Figure 6-7. Time trend of US corn crop production and yield, 1960–2005 (USDA/National Agricultural Statistical Service, 2005)

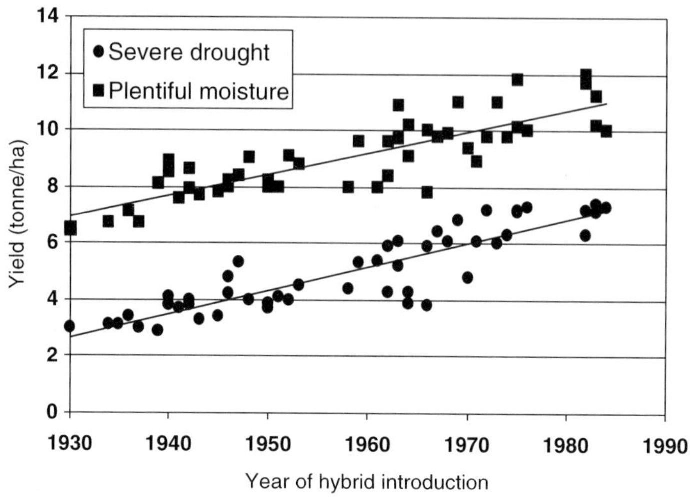

Figure 6-8. Yield of hybrid corn in wet and dry years arranged from the oldest varieties on the left to the newest on the right (Council on Agricultural Science and Technology (CAST), 1992)

Another example is to be found in Figure 6-9 (Rosenberg 1982) which shows the boundaries of the hard red winter wheat crop-growing zone in the Great Plains in 1920, 1980, and 2000. Winter wheat enjoys certain agronomic and management advantages over spring wheat which is still dominant in the northern US Plains and Prairie Provinces. Winter wheat is planted in fall, goes dormant after killing frost, breaks dormancy in spring and can be harvested from early June in Texas to September in the Prairie Provinces—early enough in much of the region to avoid the stress of midsummer heat and drought. Its relatively long growing season allows more time for photosynthetic production of sugars to be stored as starch in the grain. Spring wheat, on the other hand, must be planted after the soil has warmed sufficiently, and planting is often delayed by wet conditions in this season. Its short season allows less time for production and storage of photosynthate. Rosenberg (1982) explained that genetic and management improvements made between 1920 and 1980 allowed this expansion of the winter wheat zone. In 1980 winter wheat was growing at its northern boundary with 20% less precipitation and a 10-day shorter growing season than in 1920. The southward extension, of course, brought the crop into a hotter climate. The figure also shows that winter wheat culture continued to expand from 1980 to 1999 with expansion, indeed, in every direction.[23]

Concern is often voiced for a slowing of the rate of yield increase which has been achieved largely through traditional plant breeding. Evidence summarized

[23] I am indebted to my friend and colleague, Professor William E. Easterling, III of Pennsylvania State University for the updating of this figure from 1980 to 2000.

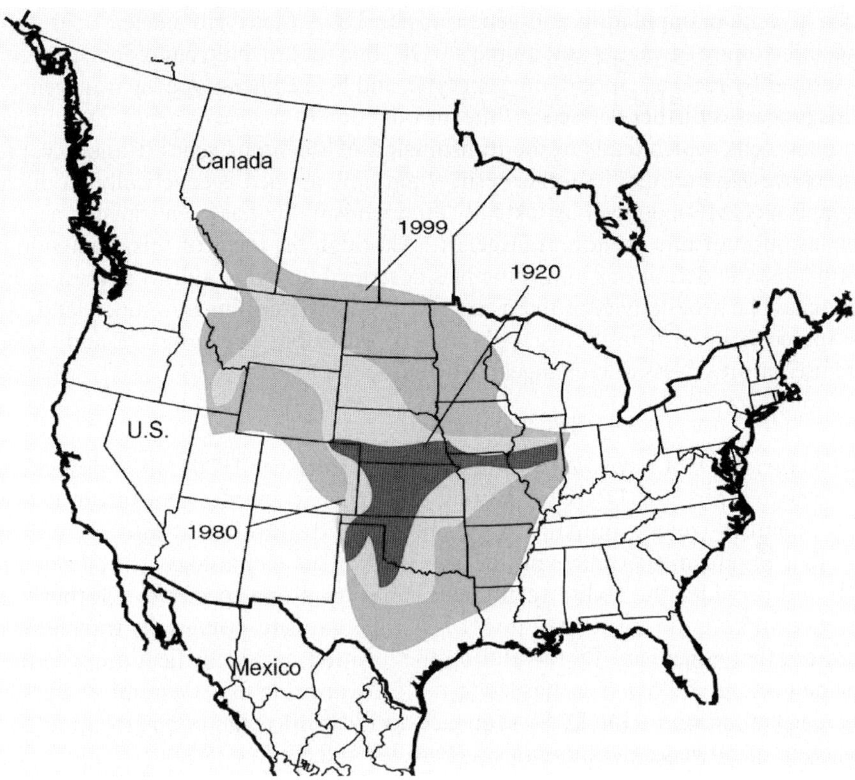

Figure 6-9. Spread of winter wheat culture from 1920 to 1999. Figure covering 1920 and 1980 from Rosenberg (1982); updated to 1999 by W.E. Easterling in 2004 (*See Color Plates*)

by Adams et al. (1999) indicates that, at least in the USA, production of agricultural crops continues to increase as it has since 1945 at a rate of ~2% per annum. Biotechnology may offer the opportunity to maintain or even increase the rate of productivity gain in crops, including those dedicated to biomass. Genetic engineering (GE), a form of biotechnology, although not without problems of various kinds, is a somewhat more efficient process than those described above for introducing new and desirable traits into plant as well as animal species. There is also a role for GE in improving the efficiency of conversion of raw biomass into transportation fuels and other products.

5.2. What is biotechnology?

In its broadest sense, *biotechnology* is the use of organisms and biological processes to provide food, chemicals, and services to meet human needs. More specifically, biotechnology has been defined by the Food and Agriculture Organization of the United Nations (FAO 2002) as "a range of different molecular techniques such as gene manipulation and gene transfer, DNA (deoxyribonucleic acid) typing and cloning of plants and animals." GE, one tool of biotechnology, has been described by FAO as "modifying genotype, and hence phenotype, by transgenes." This process is further defined as "the introduction of a gene or genes into animal or plant cells, which leads to the transmission of the input genes (transgenes) to successive generations." Another term commonly applied is genetically modified organism (GMO), defined by FAO as "an organism that has been transformed by the insertion of one or more transgenes." Although the focus of this discussion is on plants, the definitions and principles described apply as well to domesticated animals. The productivity of these animals will also determine how much land can be spared from farm and range for biomass production. The terms GE and GMO are used more or less interchangeably hereafter.

5.2.1. How are genes actually transferred?

Recombinant DNA technology, in development since the 1970s, allows scientists to remove a piece of DNA containing one or more specific genes from nearly any organism (plant, animal, bacteria, virus) and introduce it into the cell of another organism (Perseley and Siedow 1999). This technology is the basis for GE and provides the ability to transfer genes from one species to another—in essence making the entire gene pool of living organisms potentially transferable into another organism. By facilitating the location and identification of genes, the new science of *genomics* has led to rapid advances in GE. Genomics refers to the determination of the DNA sequence and identification of the location and function of all genes in an organism (Perseley and Siedow 1999).

Perseley and Siedow describe the two processes by which desirable genes are transferred into target organisms:

> In plants known as dicots (broad leafed plants such as soybean, tomato, cotton), transformation is usually brought about by the use of a bacterium, *Agrobacterium tumefaciens. Agrobacterium* naturally infects a wide range of plants and it does so by inserting some of its own DNA directly into the DNA of plants. By taking out the undesired traits associated with *Agrobacterium* infection and inserting a gene or genes of interest into the *Agrobacterium* DNA that will ultimately be incorporated into the plant's DNA, any desired gene can be transferred into a dicot's DNA following bacterial infection. The cells containing the new genes subsequently can be identified and grown using plant cell culture technology into a whole plant that now contains the new transgenes incorporated into its DNA. Plants known as monocots (grass species such as maize, wheat, rice) are not readily infected by *Agrobacterium*[24] so the external DNA that is to be

[24] Carpenter and Gianessi (2001) report an exception in that *A. tumefaciens* has been used as a vector in corn, a monocot.

transferred into the plant's genome is coated on the surface of small tungsten balls and the balls are physically shot into plant cells. Some of the DNA comes off the balls and is incorporated into the DNA of the recipient plants. These cells can also be identified and grown into whole plants that contain the foreign DNA.

Most transfers thus far have involved single genes or, at most, a few genes at a time. The introduction of whole chromosomes into plant and animal cells is an objective of continuing GE research. For reasons of predictability and stability all new crop strains developed by GE are tested for Mendelian inheritance and dominance. Those that do not display these traits are rejected for further development.

5.2.2. *Traits introduced into agricultural crops*

GE is a somewhat more efficient process than others described above for introducing new and desirable traits into plant and animal species. Herbicide tolerance (HT), insect resistance, disease resistance, and improved nutritional value are the traits most sought-after in what are still the early years of crop GE. Of these traits, genetically engineered HT and insect resistance have, thus far, experienced the most rapid rates of adoption and have had the greatest impact on agricultural markets. It is likely that efforts to breed these traits into biomass crops will also be a matter of high priority.

5.2.2.1. Herbicide tolerance HT has been bred into crop plants by conventional breeding, mutation breeding and, most recently, by GE. The trait has been engineered into soybeans, canola, cotton, corn, sugar beet, rice, flax, and other crops and is being engineered into tomato, lettuce, potato, alfalfa, wheat, and sugarcane (Gianessi et al. 2002). "Roundup-Ready" cultivar of soybean is the biggest "HT success story" so far. Roundup and other proprietary herbicides contain the active nonselective ingredient *glyphosate* which is effective against both dicot and monocot weeds. Soybeans and other species have been engineered to tolerate the glyphosate so that when it is applied early in the growing season all plants other than the herbicide tolerant ones are killed. The use of herbicide tolerant cultivars results in a marked reduction in herbicide usage as well as significant labor savings associated with a sharp reduction in the frequency of sprayings required during the course of the growing season. Other benefits attributed to glyphosate are that, compared with many of the other chemical herbicides in use today, it is much less toxic to people and animals; its half-life of 60 days is much shorter than that of other herbicides; it is more tightly bound to soil particles, thereby decreasing the danger of runoff into streams or leaching into groundwater. Crops have also been bred to tolerate a range of other chemical herbicides among which *glufosinate* produced by Liberty Link is prominent.

5.2.2.2. Insect resistance The organism *Bacillus thuringiensis* (Bt) is known to have insecticidal properties. In fact, since the 1920s Bt has been applied to crop surfaces to protect against insect attack (and also to protect against frost damage[25]). There are many strains of Bt, each lethal to specific insects. The Bt genes that confer insect resistance have been identified and transferred into a number of crop species, most notably field corn (maize) to protect against the European corn borer and cotton to protect against lepidopteron pests such as bollworm and tobacco budworm. Insect protection has also been engineered into potato and sweet corn and research is underway on peanut, broccoli, soybean, and eggplant. As with HT crops, the major benefits derived from Bt come from a reduction in the amounts of pesticide required to protect the crop and from the reduced labor and energy costs resulting from fewer applications of pesticide during the growing season. Reduced use of pesticides reduces air and water pollution, exposure of farm workers, nontarget insects and wildlife to toxic materials.

5.2.2.3. Disease resistance GE crops are also being bred for disease resistance. Virus resistance has been engineered into varieties of papaya, squash and potato, wheat, tomato, and peanut (Gianessi et al. 2002). As of 2001 there were no genetically engineered bacterial-resistant crops, although work was underway to impart such resistance to the apple, grape, and citrus. Similarly, fungal-resistant GE cultivars of potato, sunflower, and barley are in development. In addition, resistance to nematodes is being studied for pineapple and strawberry.

5.2.2.4. Nutrition and pharmaceutical properties GE is also being used to improve crop quality and nutritional characteristics of food crops. Rice is the staple food in much of Asia, Africa, and the Middle East. Vitamin A deficiency is common in these regions and GE has been applied to alleviate this problem. "Golden Rice," is a variety in which genes from the daffodil plant and a bacterial gene were engineered to produce β-carotene, a precursor of vitamin A, in its endosperm (Toenniessen 2000). Additionally, many specialists in biotechnology assert that genes producing specific compounds of pharmaceutical and immunological value can be embedded in plants that gain the ability to produce these compounds, eventually perhaps in commercially meaningful quantities. One example is the use of the potato plant to produce the hepatitis B surface antigen HbsAg (Kong et al. 2001).

Nutritional and pharmaceutical applications are probably of little relevance to GE for biomass production. Cell structural and compositional traits and overall plant "architecture" are more important and are discussed below.

5.3. Adoption of genetically engineered crops

GE cultivars first entered the scene in around 1996. In their first decade, HT has been the dominant trait, followed by insect resistance and stacked genes

[25] In this case the bacteria act as condensation nuclei causing the release of latent heat and, for a time, protecting the tissue from sub-zero temperatures and freezing.

for the two traits. The rate of adoption of biotech crops has been little short of phenomenal. In 1996, the global area planted to transgenic crops was only 1.7 million hectares. According to a report of the International Service for the Acquisition of Agri-biotech Applications (ISAAA) by James (2005) biotech crops were planted in more than 90 million hectares (222 million acres) globally in 2005, up by 9 million hectares or 11% from the previous year. In 2004 the global area of biotech crops had grown by 13.3 million hectares—up 20% from the previous year. In the USA, farmers planted 49.8 million hectares of biotech crops in 2005 (55.3% of the global area) of which ~20% were "stacked" products containing two or three bioengineered genes.

GE crops grown in the USA in 2005 were, in order of area planted, soybean, corn (maize), cotton, canola, squash, and papaya. The area of GE crops in Canada was 5.8 million hectares devoted, in order of area planted, to canola, maize, and soybean.

The ISAAA study reported that ~8.5 million farmers in 21 countries planted biotech crops in 2005, up by 0.25 million farmers and with the addition of four countries from the prior year. Notably, 90% of these farmers were in developing countries. The absolute growth in biotech crop area continues to be greater in developing countries than in the industrialized countries (6.1 million hectares). The countries accounting for the majority of the global total of biotech crop area are the USA with 55.3% of the global total; Argentina, 19%; Brazil, 10.4%; Canada, 6.4%; China, 3.7%; Paraguay, 2%; India, 1.4%; and South Africa, 0.6%. As of 2005 the dominant biotech crops were, in millions of hectares: soybean, 54.4 ; maize, 21.2 ; cotton, 9.8; and canola, 4.6.

5.4. Environmental risks of genetically engineered crops

The remarkable rate of adoption of GM crops, described above, and the prospect that the areas planted to these crops will continue to expand have not met with universal approval. Concerns have been raised that the use of GM crops may have unintended and disruptive environmental consequences. The major concerns are that: (1) individual plants of weed species growing in fields of herbicide tolerant crops will survive herbicide application and convey their resistance to future generations as well as to closely related species of plants; (2) similarly, certain target insects exposed to the *Bt* gene will develop resistance and transmit that resistance to future generations; (3) the genetic makeup of the non-GM cultivars, landraces, and related wild species will be polluted by pollen from GM crops grown in their vicinity. It is feared in all these cases that the transgenes incorporated in the genomes of these "bystander" plants will be transmitted to future generations. There is also fear (4) that the use of GM crops will contribute to the general loss of biological diversity. A further concern (5) is that humans will be directly harmed by allergenic proteins transferred into food crops. Of these concerns (1) through (3) are most relevant to the matter of GM biomass crops and are discussed below. With regard to issue (4): since biomass crops, whether herbaceous or woody, are likely to be grown, as agricultural crops are, in monoculture, they will be pose no greater risks for biodiversity than do other monocultures. In fact, since they are to be grown as perennials with a generally lesser use of pesticides, they could prove to be more hospitable to insects and animals. And, as biomass crops are not intended for direct human consumption, issue (5) does not appear relevant.

With regard to the notion of "super weeds": it is clearly possible that individual weed plants growing in fields of herbicide tolerant crops will themselves manifest tolerance to the broad spectrum herbicides used in conjunction with HT crops and that these surviving weeds will propagate future generations of HT weeds. While this threat seems ominous, it appears not different than what is already occurring in weeds grown in fields of non-GM crops subject to repeated applications of any of the hundreds of herbicides in current use.

With regard to the notion of "super-insects": the widespread use of genetically modified Bt corn, cotton, and other crops might well lead to the development of Bt resistance in both target and nontarget insects. The tactic most often proposed to avoid this effect is the establishment of *refugia* of non-Bt varieties within and adjacent to fields planted in transgenic crops. The small numbers of insects that survive contact with the Bt crops are most likely to mate with the larger populations from the refugia, thereby reducing the probability that strongly resistant progeny will appear in subsequent generations (Perseley and Siedow 1999). There is also concern that exposure to the Bt protein might affect the health of other fauna, such as birds, that ingest green tissue or seeds of the Bt plants. Although this issue is not yet fully resolved, one comprehensive study by O'Callaghan et al. (2005) reports that extensive testing on nontarget plant-feeding insects and beneficial species that has accompanied the long-term and wide-scale use of Bt plants has not detected significant adverse effects and that such plants appear to have little impact on soil biota such as earthworms, collembolans,[26] and general soil microflora.

Unintended transgene transfer is probably the most serious of the risks associated with GE crops. Pollen from transgenic crops will almost certainly be carried by wind and insect vectors to non-GE varieties of the same species, to undomesticated relatives of the transgenic crop species and to closely related species growing in adjacent fields, thereby transmitting such traits as HT, Bt, or others. This issue has been studied extensively. Eastham and Sweet (2002) assessed the risks for European agriculture and agree that unwanted crossings are likely to occur, but that risks vary greatly with species. For example, oilseed rape (Canola) is at high risk in Europe for crop-to-crop gene flow and for crop-to-wild relative gene flow. Maize is at medium risk for crop-to-crop gene flow but maize has no wild relatives in Europe although it certainly does in the Americas. Pollination is likely to occur at greater distances than the 200 m spacing recommended for isolation of GE

[26] minute wingless primitive insects

crops. Wheat is considered low risk for crop-to-crop gene flow and from crop-to-wild relative. Cross-pollination under field conditions involves less that 2% of the florets so that out-crossing generally occurs with adjacent plants. Hybrids formed by wheat and other grasses are generally sterile. Barley is at low risk because it is almost always self-fertilized or crosses only with closely adjacent plants. Clearly, containment of gene flow will not be simple. The same is true, of course, for new non-GM cultivars as well, especially those achieved through mutation breeding.

5.5. Environmental benefits of genetically engineered crops

The risks associated with the use of GM crops coupled with their wide and rapid spread in important agricultural regions of the developed and developing world has, not without cause, led to concern among many in the environmental community. But, by the same token, it is necessary to recognize that in addition to risks, GM crops also offer important environmental benefits. The benefits stem from (1) large reductions in the quantities of pesticides applied to crops and in the numbers of applications required and (2) promotion of the conversion of lands from conventional management to minimum and no-till farming—practices that come under the rubric of "conservation tillage."

With regard to pesticides: Gianessi et al. (2002) concluded on the basis of an analysis of 40 case studies in the USA that in 2001 overall pesticide use was reduced by some 20,000 tonnes of active ingredient. The greatest reduction, ~13,000 tonnes was seen in HT soybeans and ~3,000 tonnes in HT cotton.

Under conventional management, soil is plowed after harvest to incorporate residues and to prepare a seedbed for the crop to follow. That practice exposes SOM to oxidation, leading to a net emission of CO_2 to the atmosphere and to pulverization and drying of the soil surface that increase the risk of water and wind erosion. Under no-till management the soil is left undisturbed from harvest to planting. Residues remain on the surface to protect the soil from erosion and the next crop is seeded directly into a narrow seedbed or slot opened in the soil. Weed control may require herbicides to provide the crop with a competitive advantage over weeds. Although numbers are harder to come by, it appears that GM crops are fostering the adoption of no-till.[27]

5.6. Genetic engineering of biomass crops

A review of the literature indicates that research underway in government and private organizations to improve biomass crops by GE is much more modest in scope than that for agricultural crops. As indicated above, the USDOE's Biofuels Feedstock Development Program has focused most attention on switchgrass

---

[27] http://www.sagpya.mecon.gov.ar/new/0-0/agricultura/otros/granos/soja.php

(*P. virgatum* L.) and poplar (*Populus*). Research is underway on several other herbaceous species of which alfalfa (*M. sativa* L.) is notable and on other tree species including the eucalyptus (*Eucalyptus globules*), sycamore (*Acer pseudoplantanus*), sweet gum (*Liquidambar styraciflua*), black locust (*R. pseudoacacia*), and willow (*Salix caprea*).

Traditionally forage crops have been bred for yield, but even more for their digestibility as feed for ruminant animals. However, the greatest interest in biomass breeding of herbaceous plants at this time relates not to their value as animal feed but rather to their cell wall morphology and chemical composition, since these factors determine their suitability as feedstocks for ethanol production and/or for their direct use as combustible fuels (Vogel and Jung 2001). And while agricultural crops are being altered by GE to achieve HT, or insect and disease resistance, and not yet to any important degree to increase yields, switchgrass and poplar are being altered to reduce lignin content and increase cellulose and hemicellulose content (Dinus et al. 2001). In the case of poplar, GE also aims to increase wood density as a means of increasing the quantity of desirable feedstock substances extractable from a given volume of the harvested tree (Dinus et al. 2001). Poplars and other species are also being engineered for pesticide resistance, HT, and delayed flowering. Indeed, the USDA by 2002 had received applications to field-test 138 types of transformed trees (Mann and Plummer 2002).

It is also important to recognize that the objective of improving the characteristics of switchgrass, poplar, and other species for biomass continues to be met by traditional breeding programs, not only by GE. The DOE and USDA support such programs that aim to produce superior varieties of switchgrass (e.g., Vogel and Jung 2001).

Wright and Tuskan (1997) reported on the development of hybrid poplars for different regions in the USA. By that time 20 new clones of poplar had been introduced to production in the Pacific Northwest region, for example. GE is more advanced in the tree-for-biomass than in the grasses-for-biomass arena. However genetic information to develop switchgrass hybrids as well as the first molecular genetic markers for this species have been identified (Vogel and Masters 1998; Tuskan et al. 2004).

In the case of agricultural crops, protection against well-understood causes of yield loss is the major aim of transgenic varieties that have proven economically attractive thus far. In the case of biomass crops, however, desired traits are not yet fully defined because the industrial conversion processes are in early stages of development and, hence, the exact chemical composition and other desirable characteristics of biomass feedstocks are not yet known. Therefore, GE as well as traditional breeding is hampered by the lack of specific information on those traits to breed for.

Another interesting distinction relates to duration of vegetative growth. Most of the agricultural crops grown in the temperate regions are annuals and are bred so that their grains or other marketable organs mature before serious damage is done by frost. In the case of forage crops, however, an extended growing season is desirable so that more time is available for production of edible stems and leaves. The duration of vegetative growth is controlled by photoperiod. According the Vogel and Jung (2001) genetic populations of switchgrass and other herbaceous species can be moved as much as 500 km northward (in the Northern Hemisphere) where the greater daylength in summer promotes vegetative growth. Thus, if photoperiod sensitivity can be altered in productive varieties from the more southern regions, they can be grown to produce more total biomass under the longer days of the northern climes.

6. PROCESSING BIOMASS CROPS

6.1. Introduction

Biomass can be combusted directly as a boiler fuel. Oil seeds can be processed to make "biodiesel," a substitute for the petroleum-based product. Starch extracted from coarse and fine grains can be converted directly into ethanol and by-products. And plant residues, as well as grasses and woody biomass crops, can be processed through additional steps for conversion into ethanol, methanol, hydrogen, plastics, and other products. Special processing plants are required for these purposes and these are briefly described below.

6.2. Specific uses of biomass

6.2.1. Generating electricity

There are, according to the Energy Efficiency and Renewable Energy (EERE) division of the USDOE, four primary classes of biomass power systems: direct-fired, cofired, gasification, and modular systems.[28] All are used in boilers to produce steam under high pressure that drives turbines connected to electric generators. Plants powered by direct-fired biomass tend to be small compared with coal-fired plants and less efficient, converting only ~20% of their embodied energy into electric power.

Cofiring involves substitution of biomass for a portion of the coal used in existing power plants. Major modifications of the plant are not necessary although minor adjustments may be made to accommodate the biomass portion of the total fuel mix. The use of biomass lowers the total emissions of pollutants such as sulfur dioxide and nitrous oxides as well as those of heavy metals like mercury. Little is lost in efficiency and 33 – 37% of the energy in biomass is converted into electric power in cofired plants.

Solid biomass can also be converted into a gaseous form. The gas can then be run through "combined-cycle" gas turbines in a coal-fired power plant or for use

[28] http://www.eere.energy.gov/biomass/electrical_power.html

in conjunction with fuel cells. Fuel cells convert hydrogen gas to electricity and heat through an electrochemical process. Emissions from such systems should be very small and constituted primarily of water vapor.

Modular systems employ some of the same technologies but on a smaller scale for villages, farms, and small industries. EERE suggests that such systems will be useful where electricity is scarce and biomass abundant as in certain developing countries.

The USDOE's Energy Information Administration (EIA) Annual Energy Outlook for 2006 shows that in 2004 biomass accounted for ~37 billion kilowatt hours (bkWh) of US electricity generation out of a total of nearly 90 bkWh from nonhydroelectric renewable energy sources. The EIA projects that by 2030 these renewables (including biomass, geothermal, solar, wind, and municipal solid wastes will provide ~250 bkWh of which ~100 bkWh (or 40%) will be from biomass.

These forecasts also consider the state of technological change. Assuming constant technology (cost and performance of generators using renewable resources remain unchanged) total (nonhydro) renewables will provide ~ 230—100 bkWh (or 43%) from biomass. Under a high renewables scenario which assumes cost reductions of 10% in all renewable electrical generating plants, nonhydro renewables will provide ~340 bkWh of which 175 bkWh (or 51%) is from biomass.

6.2.2. *Biodiesel*

In another type of processing facility of growing importance, soybeans and other oil seeds including canola (rapeseed), mustard, and palm oil are processed to extract their oil for use as diesel fuel. "Biodiesel" is defined as "a fuel comprised of monoalkyl esters of long chain fatty acids derived from vegetable oils or animal fats." Biodiesel can be used alone or in a blend with petroleum-based diesel fuel. In North America biodiesel is produced largely from soybeans, corn, canola, cottonseed, and sunflower. US production of biodiesel was expected to triple from about 95 million liters in 2004 to about 284 million liters in 2005.[29] As of August 2005, biodiesel was produced in 35 plants in the "lower 48" of which 5 are within the Great Plains boundaries. Outside the Plains boundary, Texas has another six plants. Again as of August 2005, an additional 44 plants, four on the margins of the Plains, had been proposed for construction.[30]

6.2.3. *Ethanol from grain*

In the USA in 2005 there were some 109 fuel ethanol plants producing about 15.9 billion liters of ethanol annually. Thirty-five new plants and expansions of existing plants will increase capacity to 22 billion liters. These plants are concentrated in the US Midwest close to the reliable sources of corn, grain sorghum,

---

[29] The National Biodiesel Board, http://www.eco-web.com/register/04100.html

[30] http://agproducts.unl.edu/Biodiesel %20plant%20considerations

and wheat. Ethanol is produced in much larger quantities in Brazil using sugarcane as the raw material.

According to the Renewable Fuels Association, the Great Plans states have 42 of these plants, all but a few using grain corn exclusively for raw material. In addition to corn, milo (grain sorghum: *S. bicolor Moench*. L) is processed in two plants in Kansas, and one each in Nebraska and Texas. Barley is processed into ethanol at one plant in North Dakota. Colorado has one plant (the Coors Brewing Company in Golden) that makes 5.7 million liters of ethanol from waste beer. At this writing plants in existence or development number 13 in Nebraska, 8 in Kansas, 12 in South Dakota, 3 in North Dakota, 2 in Colorado and 1 each in Wyoming, New Mexico, and Texas. Current production in the Great Plains states is ~4.4 billion liters per annum—about one fourth of current US capacity. As of late 2005, another 1.5 billion liters per annum capacity was coming on line, again about a fourth of US capacity expansion.[31]

In 2001 Canada had six ethanol production plants in operation, of which three are in the Plains.[32] All three used wheat as their raw material and their combined production was 48 million liters per annum. Canada-wide production was then 238 million liters per annum. New facilities at that time were planned in Ontario and Quebec and were to increase national production by an additional 366 million liters per annum. The eastern Canadian ethanol plants use corn as their raw material. Canada's Ethanol Expansion Program has recently provided an additional $46 million to construct or expand five plants across Canada. Together with other existing plants these are expected to bring Canada's ethanol production up to 1.4 billion liters in 2007.[33] Two of these plants are located within the Great Plains boundaries, one in Manitoba and one in Alberta.

6.2.4. Ethanol and other products from Ligno-cellulosic materials

As explained in foregoing sections of this book, the potential for ethanol production increases greatly when lignocellulosic portions of crop plants (e.g., corn stover, wheat straw) and dedicated biomass crops can be used as the raw material. But these materials are not easily converted to sugar and require that the substrate be subjected to a "cracking" process analogous to that by which petroleum is converted into gasoline.

Lasure and Zhang (2004) describe the notion of the "biorefinery" in which renewable biomass is cracked into useful components using bioconversion technology and the resulting components are separated into useful streams for production of fuels, power and products. Corn stover can become a major source of biomass to support a lignocellulosic biorefinery. Leaving 40–60% of the stover on the field to prevent erosion, it is still possible to harvest between 54 and 91 million

---

[31] http://www.ethanolrfa.org/industry/location/

[32] Canadian Renewable Futures Association

[33] http://www.nrcan.gc.ca/media/newsreleases/2005/200550_e.htm

tonnes per annum. Estimates by Perlack et al. (2005) of the potential sustainable supply of wood waste and dedicated biomass crops suitable for this processing were given above.

Shinnar and Citro (2006) propose that, in addition to ethanol, biomass can be used to generate 'syngas' from which methanol and/or other liquid hydrocarbons can be synthesized. The syngas can be made from hydrogen gas and carbon monoxide or carbon dioxide with the H_2 generated on location by electrolysis. The oxygen coproduced in electrolysis can be used to partially oxidize the biomass. According to their calculations, this method should produce three to four times as many hydrocarbons as does fermentation to alcohol.

6.3. Bioconversion and biorefineries

All forms of biomass have the same major components—cellulose, hemicellulose, and lignin (Lasure and Zhang 2004). Cellulose is the largest fraction (40–50%) composed primarily of the 6-carbon sugar glucose. Hemicellulose is next (20–30%). Hemicellulose is a complex of primarily 5-carbon sugars, mostly xylose and arabinose. Lignin is usually 15–20% of biomass. Lignin, which provides strength to the plant structure, is a complex based on benzene rings.

Bioconversion is the use of biological processes to transform biomass materials from one form to another. Enzymes, microbes or other biological agents are used alone or in combination to accomplish bioconversions. Glucose from corn is the material from which a number of products including ethanol are made in the current "biorefinery." Conversion of lignocellulose under current technology also involves milling the biomass to produce glucose.

According to Lasure and Zhang (2004), the biorefinery of the future will be a facility for converting lignocellulosic biomass into a range of useful products—a processing unit that refines biomass. The raw materials will be "cracked" into separate components, each of which is then converted to a separate product. Processing will involve not only bioconversion but also chemical and physical cracking technologies. Lignocellulosic biomass-based refineries may begin with ethanol as their primary product. Cellulose is a glucose polymer, so that production of ethanol from glucose would likely be the first product. Efforts are also underway to engineer yeast or bacteria to convert the xylose and arabinose to ethanol.

Lasure and Zhang (2004) see two alternative patterns for the biorefinery of the future: in one case the emphasis is on maximal conversion of raw biomass to ethanol; in the other case the hemicellulose and lignin streams are converted to a wide range of value-added products. Until now, the only significant use of lignin is as a combustion fuel for power generation.

As this is written (August 2006) there are no commercial scale "biorefineries" in operation in the USA. Cargill operates a plant in Blair, Nebraska, that produces polylactic acid, a biodegradable thermoplastic, from glucose derived from starch.[34] Cargill and Natureworks are both looking at the potential of converting cellulosic biomass to ethanol and higher value products and are in various stages of research at the "bench scale" at this point.

IOGEN, a Canadian firm, operates a small pilot-plant in Ontario at which wheat straw is converted to ethanol through a microbiological process. This plant was constructed with support from the Government of Canada and Shell Canada Ltd. IOGEN has proposed to establish one plant in the USA provided some government financial aid (such as a loan guarantee) is available. This plant is currently slated to be located in Idaho. The degree of success of this facility will determine if other facilities are built.

In addition, the USDOE has issued a solicitation allotting $160 million over 3 years ($53 million in fiscal year 2007) for the construction of new biorefineries to produce ethanol from cellulosic material. These plants will have to be able to process at least 700 dry tonnes of biomass per day. The solicitation closing date was August 10, 2006. Awards will probably have been made before publication of this book.

A discussion of the advances in biochemistry and microbiology that will be needed to make bioconversion of lignocellulosic materials cost-efficient is beyond the capacity of the writer and beyond the scope of this book. It is important to note here that at this time capital costs for cellulosic conversion plants are about 4 times greater than for corn-to-ethanol and that the cost of enzymes is 10–15 times greater for cellulosic conversion.* Suffice to say here that specialists in this field are nonetheless optimistic that the diversity of microbes (most of which are yet undiscovered or uncharacterized), growing knowledge of the vast number of biological processes by which lignocellulosic materials are naturally recycled, and the new tools of genomics, proteomics, and metabolomics will lead to discovery or synthesis of microorganisms and enzymes to improve bioconversion of biomass and greatly reduce their costs (Metting et al. 2004).

6.4. Reliable supply streams

Clearly, collecting energy at the end of a pipeline or a hydroelectric turbine is far simpler and more cost-effective than combining grain or cutting and baling crop residues or biomass crops and hauling these to the boiler or processing plant. A "rule of thumb" for current grain-based ethanol production limits the biomass supplying region to roughly an 80 km radius of the processing plant. Grains are at least twice as dense as plant stems, so the mass of the latter that can be economically transported to processing centers is obviously much smaller.

---

[34] Mr. Zia Haq, personal communication, July 5, 2006. Mr. Haq of the US Department of the Energy Efficiency and Renewable Energy Division provided this information on the status in North America of biorefineries for conversion of cellulosic plant residues to ethanol and other products.

* U.S. Agriculture and the Emerging Bioeconomy. Presentation by Dr. Keith Collins, Chief Economist, U.S. Dept. of Agriculture. Conference on Advancing Renewable Energy: An American Rural Renaissance. St. Louis, MO, October 12, 2006.

Put another way—the supplying region for plant residues and dedicated biomass crops must be much larger or the power plants and lignocellulosic biomass processing plants more numerous or of much smaller capacity than those that currently process grain.

Alternatives to the corn grain-to-ethanol model are on-the-farm processing of biomass and its use on-the-farm, the development of local biorefineries or preprocessing on-the-farm (energy densification) to reduce the mass of material to be transported. Not much has yet been done to evaluate these options.

6.5. More environmental considerations

Estimates of the potential contributions of biomass to reducing or offsetting fossil fuel consumption vary widely. Optimists suggest that 20–30% can be offset worldwide, leading to a significant reduction in CO_2 emissions. When used directly as a boiler fuel, smokestack emissions of sulfur, nitrous oxides, and heavy metals are reduced, as well. Soil erosion is reduced on land planted to trees and grasses and quality of the runoff water is improved. Fertility can be improved by the accumulation of organic matter in soils planted to perennial grasses and SRWC. Perennial crops generally require less fertilizer and chemical pesticides than do annual crops. And, of course, they require a much smaller expenditure of energy for tillage. All of these factors contribute to agricultural sustainability.

Biomass-derived ethanol and methanol can reduce automobile pollution. An analysis by the Union of Concerned Scientists (UCS)[35] cites estimates by the US Environmental Protection Agency indicating that, as compared to gasoline-powered vehicles, the emissions of volatile organic compounds (VOCs) from the tailpipes of cars especially designed to burn pure methanol or ethanol could be reduced by 85–90% and carbon monoxide emissions by 30–90%. However, emission of nitrogen oxides, a source of acid precipitation, would not be reduced.

The UCS sees possible environmental disadvantages as well as advantages to biomass cropping. Mostly these stem from the notion that more land will be brought into cultivation with consequent increases in expenditures of energy to till, fertilize, and control pests on these lands. Concern is also expressed that biodiversity could be lost because of the destruction of species habitats, especially if forests are managed more intensively. Technological optimists (e.g., Waggoner 1994) would argue that increasing agricultural (including biomass) productivity through the use of best-management practices and genetic improvements frees land for uses other than food production, making more available for ecologically beneficial purposes. There are strong arguments and valid concerns on both sides of this question making further analysis imperative.

Another environmental concern about biomass cropping is raised by Raghu et al. (2006). They cite examples of how introduced forage grasses have become

---

[35] http://www.ucsusa.org/clean_energy/renewable_energy_basics/environmental-impacts-of-renewable-energy-technologies.html

invasive weeds in some instances and suggest that this could also occur with Miscanthus, switchgrass, and other candidate biomass crops. These grasses could spread into surrounding fields or rangelands by seed propagation or vegetatively from the fields in which they are planted. As weeds in agricultural fields, these grasses can probably be kept under control with chemical herbicides, but it may be too expensive to do so if they spread into rangelands. Raghu et al. urge that agronomic and ecological analyses, such as are already mandatory for transgenic plants, will be needed in order to assess the invasive potentials of biofuel crops before they are introduced to new regions.

6.6. Energetics

Boosters of ethanol from grain corn tout the product as an environmentally benign substitute for gasoline. Critics argue that this product requires almost as much (or perhaps even more) energy to produce as it contains—in other words that ethanol has a near neutral or even negative net energy balance (NEB). The literature contains many reports of positive, neutral or negative NEBs for ethanol from grain, but the calculations are all very much dependant on the specific energy costs considered. Most analyses include the energy embedded in fertilizers and agricultural chemicals, energy expended in traction and transportation and processing plant operations.

The most definitive life cycle accounting of NEB for ethanol from grain and for biodiesel from soybeans is that of Hill et al. (2006). In addition to the costs cited above, they considered the energy costs required to grow the seed and to produce the farm machinery required in corn and soybean production. The energy required to construct the buildings in which machinery is built, the energy costs of building and operating biofuel production facilities, and the energy costs of maintaining production facility workers and their households are also considered. On the "gains" side the analysis the energy embedded in the ethanol or biodiesel is accounted for as well as that in by-products such as distiller's dry grain.

The news from this most comprehensive of studies is better than previously thought. Ethanol from corn grain yields 25% more energy than is invested in its production. Biodiesel from soybeans yields fully 93% more energy than is required to produce it. In addition, GHGEs are reduced 12% by the production and combustion of ethanol and 41% by biodiesel relative to the fossil fuels they replace. Further, biodiesel releases only 1%, 8.3%, and 13% as much nitrogen, phosphorous, and pesticide, respectively, than ethanol releases per net energy gain.

Unfortunately, Hill et al. do not include cellulosic ethanol in their analysis. It seems reasonable to assume that its NEB will lie somewhere between those of corn grain ethanol, and soybean biodiesel. The NEB of corn stubble—and wheat straw—to ethanol (or for direct firing) should be strongly positive since these crops are grown for their grain and the residues can be considered "free" if all energy costs are attributed to the harvested grain. Dedicated biomass crops such as switchgrass are generally less demanding of fertilizer, chemicals, and tillage and should have more positive NEBs than that of corn grain ethanol, although energy costs for transporting them will probably be higher. Processing costs could also prove to be higher.

6.7. Market penetration issues

6.7.1. Prerequisites for penetration

For biomass to penetrate the energy market will require the availability of land to produce a renewable, reliable supply to the power plants and biorefineries that will be retooled or newly built to process it and a market or supported price for bioproducts that will make and keep them competitive with fossil-based fuels and their derivative products. First we need to explore the rationale for the current emphasis on biomass and what appears to make it an attractive prospect for market penetration.

McCarl et al. (2005) cite a number of factors now motivating the development of all kinds of mitigation technologies to reduce GHGEs. In general and in particular ways these relate to biofuels. The factors they discuss are presented, almost verbatim, as follows:

- *First* is the need for precaution. Since the timing of emissions, their impacts on climate change, the economic implications, and reversibility of deleterious effects are all highly uncertain, it may be desirable to "go slow" and preserve options for mitigation and/or adaptation. Biomass is one technology offering the opportunity to reduce emissions and slow down the rate of climate change.
- *Second* is the increase of international pressures on the USA, responsible for more than a quarter of global GHG emissions, to reduce them.
- *Third* are domestic policies in the USA including the "Clear Skies" program that indicates the need for future actions to reduce GHGEs as well as emissions of sulfur and nitrous oxides, the precursors of acid rain, and of mercury, as well. Biomass would help accomplish these emissions reductions.
- *Fourth* is the realization that manufacturing and the energy industry face great uncertainty as to whether GHGEs controls will be imposed in the coming decades and require of them significant reductions in production and sales, sales which could by then be considerably larger than they were in the Kyoto Protocol[36] base year of 1990. Industries have already begun to search for economically sound ways to reduce their emissions, as required.
- *Fifth* is the prospect of a need for cheap emissions reduction options. Biofuels, as one such option, has the advantage of relatively low per ton carbon costs and

[36] The Kyoto Protocol to the United Nations Framework Convention on Climate Change assigns mandatory targets for the reduction of greenhouse gas emissions to signatory nations. The Protocol entered into force on February 16, 2005. As of July 2006 there were 164 signatory countries. Canada is a signatory and ratified the Protocol in December 2002; the United States is a signatory but, as of this writing (August 2006), does not intend to sign. Source: http://en.wikipedia.org/wiki/List_of_Kyoto_Protocol_signatories

offers the possibility that adjustments can be made relatively quickly. A number of biofuel and agricultural sequestration strategies are already in effect (e.g., corn grain-based ethanol production, minimum tillage to improve soil quality, combustion of wood processing wastes for power generation and others)—motivated by factors other than concern about changing climate. Well-known technologies can be quickly deployed—buying time for development of such complex mitigation tools as geological and ocean sequestration—or allowing certain industries to avoid such investments entirely.

- *Sixth* is the fact that biofuels and agricultural and forest-carbon sequestration practices that offset GHGEs are consistent with other governmental policies favoring practices designed to achieve both environmental improvements and agricultural income support.
- *Last, but of major political importance*, is the opportunity that carbon sequestration and biomass fuels offer another market for farm products—a market in which credits reductions or offsets for CO_2 or other GHGEs can be traded. In these markets, which are under development, offset producers could sell GHGE reductions rights.

6.7.2. *Supply*

Supply requires that an adequate land area can be devoted to production of dedicated biomass crops and/or that adequate quantities of crop residues, forest wastes, manure, and industrial by-products can be collected. Does North America have the requisite land areas? Although their estimates differ, both Lave et al. (2002) and Perlack et al. (2005) are optimistic on that score. For example, Lave et al. estimate that for biofuels to replace the 492 billion liters of gasoline used annually by the US light duty fleet of vehicles would require a renewable biomass supply from 121 to 202 million hectares of land—17–28% of the 728 million hectares land area of the lower 48 states. Most of this land, by Lave's account, is now in grassland pasture and range (238 million hectares), forest (263 million hectares), or cropland (186 million hectares).[37] The needed 121–202 million hectares could be supplied from 16 million hectares of high-productivity idled cropland, 18 million hectares of land used to grow grain now sold at below production cost and from pasture land and forests not associated with farms. Further, according to Lave et al., the requisite land area could be assembled without disturbing the nation's parks, wilderness areas, wetlands, or built areas.

Perlack et al. (2005) suggest that, when all renewable sources of biomass are considered, a much smaller area of land—22 million hectares—need be converted from agricultural to dedicated biomass cropping. These authors also hold that no major disruptions need occur.

---

[37] Estimates vary; e.g., McCarl and Schneider (ca. 2001) put the area of cropland at 132 million hectares.

Nonetheless, a major obstacle to biomass penetration can be the US public's possible resistance to the notion of altering use of so much of the nation's land resource. But many arguments of societal benefits can be brought to bear. If properly managed, Lave et al. (2002) argue, grasses and trees dedicated to biomass production would return the land more closely to its original vegetation, providing habitat for endangered species, perennial vegetative cover to protect against erosion, and other benefits. Other ecological benefits such as increased biodiversity, enhanced recreational opportunities, and the new economic and employment opportunities that would follow from the conversion of so large an area of US land could make the prospect politically feasible.

Other environmental arguments might influence public attitudes. Schneider and McCarl (2003) and McCarl et al. (2006 in press) pose four possible outcomes of expanded biofuel production: a widespread biofuel market would support agricultural prices and incomes by adding to demand, replacing other forms of farm income support; replacement of some fuel additives such as methyl tert-butyl ether (MTBE) with ethanol would be environmentally desirable; similarly, replacement of coal with biomass would reduce mercury pollution; substitution of biofuels-based products for petroleum would reduce dependence on imports and contribute to energy security and, of course, biofuel combustion would substantially offset net GHGEs by the recycling of carbon.

Another argument supporting the public (in this case the global public) benefits of a biomass economy is made by Cline (2004). He holds that a reduction in global poverty would follow a rise in agricultural prices due to displacement by biomass crops of a substantial area of land now devoted to agricultural crop production. Since the bulk of the world's poor are in the agricultural sector, they would benefit from such a rise in prices. Although Cline's point is well-taken, it does not seem likely that in the USA, Canada (or most any other country), the prospect of higher food prices would, in fact, encourage public support of conversion to a biomass economy. And, as suggested in an earlier section of this chapter, increased or rising agricultural productivity, particularly the application of GE, would likely compensate for at least some of the loss of crops due to land conversion.

6.7.3. Price of biofuels

When Lave et al. (2002) did their analysis the major obstacle to adoption of biofuels they foresaw was one of cost to the consumer. They calculated that motorists in the US market would not switch to ethanol fuel unless the price of gasoline reached ~$0.70 per liter ($2.70 per gallon). That seemed an astronomical price at the time as gasoline was then selling for ~$0.40 per liter. At the time of this writing, regular gasoline is selling close to $0.80 per liter.

A frequent criticism of the corn grain-to-ethanol enterprise has been that, to be marketable at all, the product must be subsidized. Today ethanol continues to be subsidized at $0.13–0.15 per liter. These subsidies can take three forms: a 13.5¢ per liter ($0.51 per gallon) rebate of the federal fuel tax for ethanol added to gasoline products by refiners; or a 14.2¢ per liter ($0.54 per gallon) credit to refiners off their federal income tax and a 2.6¢ per liter ($0.10 per gallon) tax credit for small producers (defined as producing up to 230 million liters).[38]

McCarl et al. (2005) have argued that continued subsidization of biofuels to make them more competitive in the market with fossil based fuels may well be justified by their environmental and geopolitical benefits. On this same note, Cline (2004) argued that the cost disadvantages facing biomass vis-à-vis petroleum-based fuels could warrant public sector intervention to increase its competitiveness if one considers the environmental damage done by each ton of carbon emitted by fossil fuel combustion and there is no equivalent tax on carbon-emitting fuel. He states: "it makes sense from a policy standpoint to provide a subsidy to carbon free alternative fuels commensurate with the damage they avoid."

Gasoline prices in the USA and Canada have risen in the past few years to levels that clearly make ethanol fuels more attractive, perhaps even competitive with gasoline. One need not be a professor of economics to foresee that the rise in demand for petroleum in China and India, political instability in oil-producing nations in the Middle East, Central Asia, and Africa and tense relations between the USA and Latin American petroleum-producing nations virtually assures that petroleum prices will not recede to their early 2000s levels and that, more likely, these prices will continue to rise. If gasoline prices remain at the $0.80 per liter level or higher and if cellulosic biomass conversion proves cheaper than ethanol-from-grain, the competitive disadvantage faced by ethanol vis-à-vis gasoline will diminish and pressure will grow to reduce or eliminate at least some of the current ethanol subsidy.

6.7.4. Carbon taxes and trading systems

A "carbon tax" on energy sources which emit CO_2 into the atmosphere has been proposed as a means of reducing GHGEs. Essentially, a carbon tax would establish a market price for carbon emitted to the atmosphere and allow those who capture, sequester or otherwise offset carbon emissions to gain from their actions.

All manner of climate change mitigation strategies have been proposed including formal emissions credits trading schemes. Studies have been made in order to estimate how high carbon taxes (or trading credits) need be in order to make these strategies economically competitive with fossil fuels. Schneider and McCarl (2003) at Texas A&M University used Forest and Agricultural Sector Optimization Model (FASOM), to estimate how the carbon tax (the dollar value attributed to each ton of carbon emissions) would influence land conversion from agriculture to biomass production in the USA. In their analysis a carbon value of $54 per tonne is required to initiate land conversion. It takes a carbon tax of $454 per tonne to encourage conversion of about 49 million hectares. Accordingly, the stock of

[38] Steven Pearlstein, "Going crazy for ethanol," *Washington Post*, May 24, 2006.

agricultural land decreases from about 138 million hectares at the \$54 carbon-tax level to about 89 million hectares at the \$454 level.

The calculated effect of these changes on agricultural prices is minor (~3.5% rise) until the \$54 per tonne carbon-tax threshold is reached. Between \$54 and \$181 per tonne average prices rise ~5% for every \$18 incremental increase in the price of carbon. Commodity prices rise because of increased competition for land and because the carbon tax increases the cost of production inputs (fuel, fertilizer, etc.). The McCarl and Schneider calculations also indicate that carbon taxes between \$54 and \$181 per tonne yield a 6% decrease in food exports for each \$18 per tonne increase in the carbon tax. In this analysis only biomass for direct combustion (boiler fuel) and ethanol grain were considered. Conversion of lignocellulosic biomass to ethanol was not considered.

The relative importance and likely phasing-in of agricultural carbon sequestration options and biomass (for power plant feedstock) was analyzed in a subsequent study by the Texas A&M group (Lee et al. 2005) using FASOMGHG, a GHG version of the FASOM model. This analysis concluded that the optimal mitigation portfolio needed to offset 3–15% of US projected GHGEs by 2010 (8–12 million metric tonnes of CO_2 equivalent) can be accomplished with a CO_2-equivalent price ranging from \$4.5 to \$45 per tonne of carbon. Agricultural soil carbon sequestration is most efficient at low prices; forest sequestration becomes more efficient at prices above ~\$9 per tonne. Since soil carbon sequestration, whether in agricultural fields or in forests, "saturates" after 40–60 years and "leakage" leads to losses of the sequestered carbon, Lee et al. think its long-term importance may be exaggerated. In this analysis power plant biomass feedstocks and afforestation become more important in the longer run and at higher carbon prices.[39]

The McCarl et al. (2005) calculations lead to the conclusion that at high carbon prices biomass feedstocks can be a way of reducing GHG emissions from US electrical generation, but appear to be of limited usefulness in the liquid fuel markets—this until such time as improved production methods for biofuel crops are developed. But what will be needed to facilitate penetration into the transportation fuel market?

Lave[40] concluded that the prerequisites for competitive lignocellulosic ethanol are: technological advances and reduced production costs, infrastructure

[39] The European Union has established a formal mechanism, the European Emissions Trading System (ETS), in effect for 2005–2007. According to McCarl (2006, unpublished report, Department of Agricultural Economics, Texas A&M University, College Station, TX 77845) prices per tonne of CO_2 equivalent in ETS fluctuated between \$10 and \$34 during spring of 2006. At the Chicago Climate Exchange, an experimental market, prices ranged between \$1 and \$3 per tonne in the spring of 2006. The McCain-Lieberman Climate Stewardship Act (proposed in 2003) encouraged establishment of a market driven greenhouse gas emissions reduction scheme. An analysis of the requirements of McCain-Lieberman yields a CO_2 equivalent price of about \$10 per tonne. Thus far the Act has failed to pass in the US Congress.

[40] L. Lave, June13, 2003. Presentation to the National Commission on Energy Policy Forum, "The Future of Biomass and Transportation Fuels" meeting Hart Senatorial Office Bldg, of the US Congress.

development or higher petroleum prices or stringent GHG emissions legislation or subsidized production (lower taxes) or consumer demand for renewable fuel, and consumer acceptance of major land use change. As all readers will know, the higher petroleum prices are already with us and likely to stay with us. Technological advances, some described in preceding sections of this chapter, appear likely. Both the high price of oil and advancing technologies may be good news from the point of view of reducing carbon emissions.

6.8. Capture and sequestration of biomass carbon

Biomass, we know, recycles carbon from the atmosphere. But what if the carbon embedded in biomass is captured at the smokestack of the power plant where it is combusted as fuel and then sequestered in geologic storage?

The USDOE has made substantial investments in research on modalities for sequestering CO_2 in geologic strata. A report on the potential for CO_2 storage in North America (defined in this case as Canada and the 48 conterminous states) by Dahowski et al. (2004) identifies 326 onshore candidate geologic reservoirs, each capable of storing at least 1 Mt of CO_2. Their combined storage capacity is ~3800 Gt CO_2. When this report was issued there were 2082 anthropogenic point sources in North America with annual emissions greater than 100,000 tonnes of CO_2. Power plants account for 66% of the emissions in North America; gas processing for 22% and refineries for 4.6%. Iron and steel, cement, ethylene, oil sands, hydrogen, ammonia, ethanol, and ethylene oxide plants account for the remaining 8%.

Figure 6-10 shows the location of geologic formations underlying the Great Plains or within 160 km (100 miles) of its boundary deemed suitable for storage of CO_2. These formations include deep saline aquifers, basalts, coal basins, gas basins, and oil plays.[41] Deep saline aquifers underlie most of the Canadian Plains, the western Dakotas, and eastern Montana. The southern Great Plains states are underlain by gas basins and oil plays and the northern Plains by coal basins. The greatest potential for CO_2 storage is in deep saline aquifers, found largely in the northern Plains states and Prairie Provinces. The eastern portions of North Dakota, South Dakota, and Nebraska lack the geological formations useful for CO_2 storage. Overall the region defined in Figure 6-10 has the capacity to store 1489 Gt CO_2, which is nearly 40% of the total North American storage capacity. The breakdown of this storage capacity is given in Table 6-7.

It is likely that most of the biomass produced on the Great Plains will be used for ethanol production (from grain or cellulosic materials) or for fuel in electricity generation. In the former case, the carbon withdrawn from the atmosphere and fixed by photosynthesis in biomass is ultimately returned to the atmosphere from

[41] Injection of CO_2 under pressure to extract petroleum from depleted oil strata has been practiced for many years.

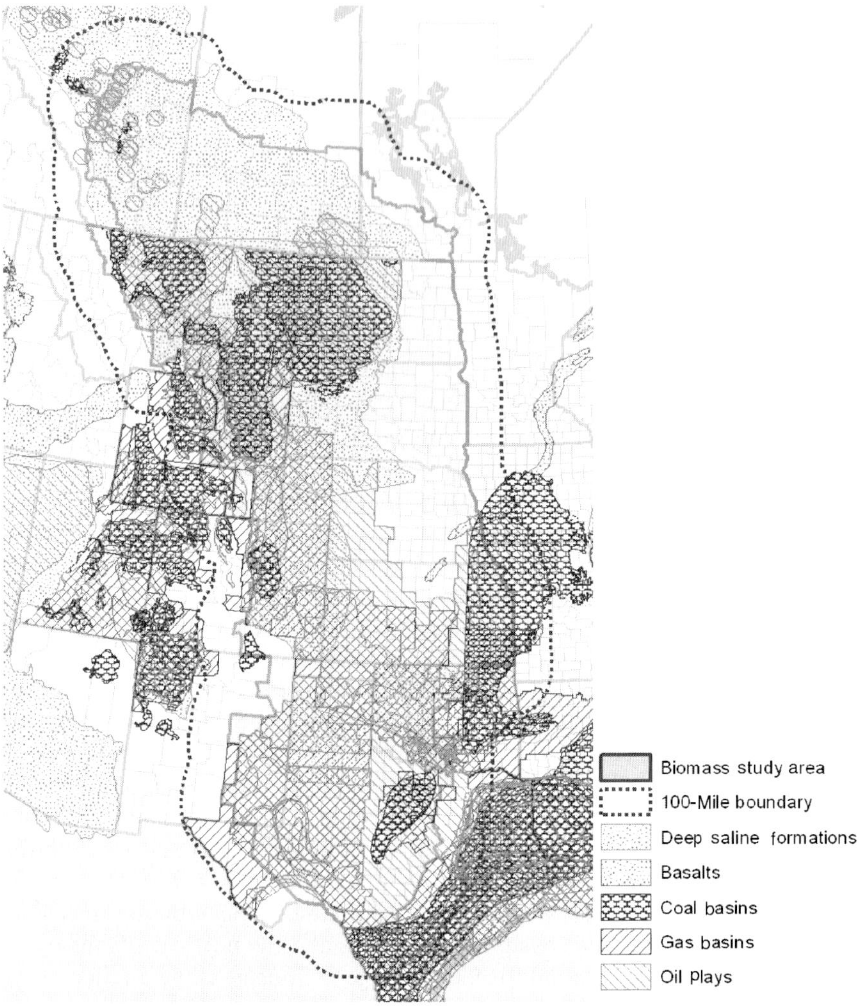

Figure 6-10. Geologic CO_2 storage potential within the North American Great Plains and the surrounding 160 km (100 miles). (Courtesy of J.J. Dooley and C.L. Davidson, Pacific Northwest National Laboratory) (*See Color Plates*)

the tail pipes of motor vehicles. In the case of electricity generation the CO_2 is emitted through smokestacks. If that CO_2 can be captured and sequestered in nearby geologic strata, the near-zero net emission of biomass CO_2 can become a means to achieve negative-emissions of this GHG, a way to actually reduce its atmospheric concentration. The coupling of biomass for energy with carbon capture and sequestration (CCS) provides an opportunity to contribute to the mitigation of potential greenhouse warming and climate change—an opportunity that the Great Plains region may be well suited to seize.

Table 6-7. CO_2 sequestration potential in geological strata within and adjacent to the North American Great Plains. (Courtesy of J.J. Dooley and C.L. Davidson, Pacific Northwest National Laboratory)

Formation	$GtCO_2$
Deep saline formations	1,425
Coal basins	33
Basalts	11
Gas basins	19
Oil plays	1

Smith et al. (preliminary report, 2006) used an integrated assessment model ObjECTS MiniCAM (Kim et al., submitted) to assess the long-term potential role of alternative energy, economy, and environmental regimes. The model has a detailed, technologically explicit energy sector with biomass fuels supplied from residue sources and from dedicated energy crops.

The Smith et al. analysis is global in scale and does not consider the Great Plains region explicitly. Nonetheless its findings should be generally applicable to the region. Five technology scenarios are considered: (1) no-CCS, (2) fossil CCS, and three biomass cases. These are (3) large oxygen-blown systems involving fossil and biomass with CO_2 removal fractions of 90% or better, (4) "expensive BioCCS" systems similar in technology but with 50% higher costs and slightly lower electric conversion efficiency, and (5) atmospheric pressure steam-blown systems with 44% CO_2 removal fraction. Of these (4), in particular, could be based on a supply of biomass from limited agricultural areas.

Smith et al. calculated total global costs of the 5-carbon policy options to stabilize atmospheric CO_2 at concentrations between 450 and 750 ppmv. Fossil fuel CCS (case 2) dramatically lowers cost from a no-CCS policy (case 1) across the range of concentrations from 450 to 750 ppmv. Biomass options (cases 3 and 4) reduce costs still further, particularly in the 450–550 ppmv stabilization range. In dollar terms carbon prices do not exceed $200 per tonne (2005 USD) except for target concentrations below 500 ppmv. At 450 ppmv the cost of biomass CCS is 40% of that for fossil CCS alone; at 650 ppmv the cost of biomass CCS is 70% that of fossil CCS alone.

One O^{bj}ECTS MiniCAM model run suggests that if the price of carbon were sufficiently high biomass might be used as a "scrubber" of CO_2 from the atmosphere by combusting it and sequestering the carbon even without using its for power generation or conversion to ethanol. Obviously, though, using the biomass first to provide value makes better economic sense. In the case of electricity generation with biomass, the price of carbon (credit for sequestration) can be passed on to the consumer in the form of lower energy rates.

6.9. Outlook

Petroleum and natural gas prices surged in 2005, driven by the rapid worldwide increase in energy consumption, especially in China and India, and the effects of hurricanes Katrina and Rita on oil production and processing on the US Gulf Coast. Petroleum prices rose above \$60 per barrel—considerably higher than the levels that many economists thought would be necessary to make biomass competitive in the energy market. After a brief decline from its peak, prices rose again (as of August of 2006) to about \$75 per barrel and have since preceded. But there is little to suggest that, in the long run, the days of cheap oil will ever return.

The political will to support a biomass option is growing in both the USA and Canada. The Energy Policy Act of 2005 (US Congress 2005) was approved with strong bipartisan support by both the US House of Representatives and the Senate and was signed into law by the president. The Act contains a large number of provisions to advance biofuels, biobased products, and the industrial biotechnology used in their production. The bill includes nearly \$3.6 billion in authorizations for biomass,bioenergy, and biorefinery research, development, and demonstration programs. It anticipates a doubling of the volume of renewable fuels in the nation's fuel supply by 2012. The bill also requires, starting in that year, the use of 250 million gallons (~945 million liters) per year of ethanol distilled from cellulosic materials. The bill provides grants, incentives, and loan guarantees for the construction of biorefineries and the production of cellulosic ethanol. The Act also expands the federal government's biobased product procurement requirements, and provides tax incentives for biofuels production and distribution. Many efforts are underway in Canada, too, to further the development of renewable energy sources, biomass prominent among them.[42]

In view of the surging petroleum, gas, and electricity prices, public interest in, and support for, biomass energy is likely to be strong in coming years. The major concerns of environmental interest groups is—and likely will continue to be—the conversion of large areas to dedicated energy crops with its implications for land use, deforestation in particular. Indeed, in the tropics—Brazil, for example—the expansion of sugarcane production and that of other agricultural crops may well increase and intensify tropical deforestation. In the Plains, on the other hand, a conversion of land use back to perennial grass cover may be ecologically benign and desirable.

Biomass culture, as compared with conventional agricultural cropping, has proven environmental benefits, such as reductions in fertilizer use, irrigation water use, and traction for tillage as well as the potential to reduce the rate of greenhouse warming. Broad public acceptance should be easy to obtain, especially if large-scale biomass production is accomplished without large concomitant increases in the costs of food and fodder.

---

[42] http://oee.nrcan.gc.ca/Publications/statistics/parliament04-05/chapter7.cfm?attr=0

REFERENCES

Adams, R.M., Hurd, B.H., and Reilly, J., 1999, *A Review of Impacts to U.S. Agricultural Resources*, Report prepared for the Pew Center on global climate change, 39 pp.

Berdahl, J.D., Frank, A.B., Krupinsky, J.M., Carr, P.M., Hanson, J.D., and Johnson, H.A., 2005, Biomass yield, phenology, and survival of diverse switchgrass cultivars and experimental strains in western North Dakota, *Agron. J.* **97**:549–555.

Brown, R.A., Rosenberg, N.J., Hays, C.J., Easterling, W.E., and Mearns, L.O., 2000, Potential production and environmental effects of switchgrass and traditional crops under current and greenhouse-altered climate in the Central United States: a simulation study, *Agric. Ecosyst. Environ.* **78**:31–47.

Carpenter, J.E. and Gianessi, L.P., 2001, *Agricultural Biotechnology Updated Benefits Estimate*. National Center for Food and Agricultural Policy (NCFAP), Washington, DC.

CAST (Council for Agricultural Science and Technology), 1992, *Preparing US Agriculture for Global Climate Change*, Task Force Report No. 119, Ames, Iowa, 96 pp.

Cole, V., Cerri, C., Mosier, A., Rosenberg, N.J., and Sauerbeck, D., 1996, Agricultural options for mitigation of greenhouse gas emissions, Chap. 23 in Watson, R.T., Zinowera, M.C., and Moss, R.H. (eds.), *Climate Change 1995: Impacts, Adaptations and Mitigation of Climate Change*, Report of IPCC Working Group II, Cambridge University Press, Cambridge, pp. 745–771.

Coleman, K. and Jenkinson, D.S., 1966, RothC-26.3—A model for the turnover of carbon in soil, in Powlson, S.S., Smith, P., and Smith, J.U. (eds.), *Evaluation of Soil Organic Matter Models Using Existing Long-Term Datasets*, NATO ASI Series I, Springer, Heidelberg, pp. 237–246.

Cline, W.R., 2004, *Biomass, Global Warming and Global Poverty*, Institute for International Economics, Washington, DC (seen in manuscript, March 2004), 25 pp.

Dahowski, R.T., Dooley, J.J., Davidson, C.L., Bachu, S., and Gupta, N., 2004, *A CO_2 Storage Supply Curve for North America*, Report PNWD-341, Battelle Pacific Northwest Division, 92 pp. Plus appendices.

Dinus, R.J., Payne, P., Sewell, M.M., Chiang, V.L., and Tuskan, G.A., 2001, Genetic modification of short rotation poplar wood: properties for ethanol fuel and fiber productions, *Critical Reviews in Plant Science* **20**:51–69.

Duffy, M.D. and Nanhou, V.Y., 2002, Costs of producing switchgrass for biomass in southern Iowa, in Janick, J. and Whipkey, A. (eds.), *Trends in New Crops and New Uses*, ASHS Press, Alexandria, Virginia, pp. 267–275.

Eastham, K., and Sweet, J., 2002, *Genetically Modified Organisms: The Significance of Gene Flow Through Pollen Transfer*, European Environmental Agency, Environmental Issue Paper No. 28, Copenhagen, Denmark, 75 pp.

Edmonds, J.A. and Clarke, J.F., 2004, Biotechnology and climate change: an introductory essay, Chap. 1 in Rosenberg, N.J., Izaurralde, R.C., and Metting, F.B. (eds.), *Applications of Biotechnology to Mitigation of Greenhouse Warming*. Proceedings of the St. Michaels II Workshop, April 2003, Battelle Press, Columbus, Ohio, pp. 1–16.

Edmonds, J.A., Wise, M., and MacCracken, C., 1994, *Advanced Energy Technologies and Climate Change: An Analysis Using the Global Change Assessment Model (GCAM)*, PNL-9798, Pacific Northwest Laboratory, Richmond, Washington.

FAO (UN Food and Agricultuure Organization), 2002, *FAO Glossary of Biotechnology for Food and Agriculture*, Biotechnology in Food and Agriculture, http://www.fao.org/biotech/find-form-n.asp

Fisher, M.J., Rao, I.M., Ayarza, M.A., Lascano, C.E., Sanz, J.I., Thomas, R.J., and Vera, R.R., 1994, Carbon storage by introduced deep-rooted grasses in the South American Savannas, *Nature* **371**:236–238.

Gebhart, D.L., Johnson, H.B., Mayeux, H.S., and Pauley, H.W., 1994, The CRP increases organic carbon, *J. Soil Water Conserv.* **49**:488–492.

Gianessi, L.P., Silvers, C.S., Sankula, S., and Carpenter, J.E., 2002, *Plant Biotechnology: Current and Potential Impact for Improving Pest Management in U.S. Agriculture*, National Center for Food and Agricultural Policy (NCFAP), Washington, DC.

Giorgi, F., Mearns, L.O., Shields, C., and McDaniel, L., 1998, Regional nested model simulations of present day and 2 × CO_2 climate over the Central Plains of the US, *Clim. Chang.* **40**:457–493.

Grant, R.F., Juma, N.G., and McGill, W.B., 1993, Simulation of carbon and nitrogen transformations in soil: mineralization, *Soil Biol. Biochem.* **25**:1317–1329.

Grant, R.F., Izaurralde, R.C., Nyborg, M., Mahli, S.S., Solberg, E.D., and Jans-Hammermeister, D., 1998, Modeling surface tillage and residue effects on soil C storage under ambient vs. elevated CO_2 and temperature in ecosystems, in Lal, R., Kimble, J., Follett, R., and Stewart, B.A. (eds.), *Soil Processes and the Carbon Cycle, Advances in Soil Science*, CRC/Lewis Publishers, Boca Raton, Florida, pp. 527–547.

Gravert, C.E., Tiffany, L.H., and Munkvold, G.P., 2000, Outbreak of smut caused by *Tilletia maclaganii* on cultivated switchgrass in Iowa, *Plant Health Progress*, doi:10.1094/PHP-2000-0815-01-HN.

Hill, J., Nelson, E., Tilman, D., Polasky, S., and Tiffany, D., 2006, *Environmental, Economic, and Energetic Costs and Benefits of Biodiesel and Ethanol Biofuels*, Proceedings of the National Academy of Sciences (early edition), 5 pp., www.pnas.org/cgi/doi/10.1073/pnas.0604600103

IPCC (Intergovernmental Panel on Climate Change), 1996, *Climate Change 1995: The Science of Climate Change: Report of Working Group I*, Houghton, J.T., Meira Filho, L.G., Callander, B.A., Harris, N., Kattenberg, A., and Maskell, K. (eds.), Cambridge University Press, New York, p. 4.

IPCC (Intergovernmental Panel on Climate Change), 2000, *Special Report on Emissions Scenarios*, Nakicenovic, N. and Swart, R. (eds.), Cambridge University Press, Cambridge, 612 pp.

Izaurralde, R.C., Rosenberg, N.J., and Lal, R., 2001, Mitigation of climatic change by soil carbon sequestration: issues of science, monitoring and degraded lands, *Adv. Agron.* **70**:1–75.

Izaurralde, R.C., Williams, J.R., McGill, W.B., Rosenberg, N.J., and Quiroga-Jakas, M.C., 2006, Simulating soil C dynamics with EPIC: model description and testing against long-term data. *Ecological Modeling* **192**:362–384.

Jain, A.K., West, T.O., Yang, X., and Post, W.M., 2005, Assessing the impact of changes in climate and CO_2 on potential carbon sequestration in agricultural soils, *Geophys. Res. Lett.* **32**, L19711, doi 10.1029/2005GL023922.

James. C., 2005, *Global Status of Commercialized Biotech/GM Crops: 2005*, ISAAA Brief 34-2005, http://www.isaaa.org/kc/bin/briefs34/es/index.htm

Kim, S.H., Edmonds, J.A., Smith, S.J., Wise, M., and Lurz, J., 2006, The O[bj]ECTS framework for integrated assessment: Hybrid modeling of transportation. *Energy J.* (in press)

Kiniry, J.R., Williams, J.R., Gassman, P.W., and Debaeke, P., 1992, A general, process-oriented model for two competing plant species, *Trans. Amer. Soc. Agric. Eng.* **35**:801–810.

Kong, Q.X., Richter, K.L., Yu, F.Y., Arntzen, C.J., Mason, H.S., and Thanavala, Y., 2001, Oral immunization with hepatitis B surface antigen expressed in transgenic plants. *Proc. Natl. Acad. Sci. USA* **98**:11539–11544.

Lasure, L.L. and Zhang, M., 2004, Bioconversion and biorefineries of the future, in Rosenberg, N.J., Izaurralde, R.C., and Metting, F.B. (eds.), *Applications of Biotechnology to Mitigation of Greenhouse Warming*, Proceedings of the St. Michaels II Workshop, April 2003, Battelle Press, Columbus, Ohio, pp. 95–109.

Lave, L.B., Griffin, W.M., and McLean, H., 2002, The ethanol answer to carbon emissions. *Issues in Science and Technology*, Winter 2001–2002, pp. 73–78.

Lee, D.K. and Boe, A., 2005, Biomass production of switchgrass in central South Dakota, *Crop Sci.* **45**:2583–2590.

Lee, H.-C., McCarl, B.A., and Gillig, D., 2005, The dynamic competitiveness of US agricultural and forest carbon sequestration, *Can. J. Agr. Econ.* **53**:343–357.

Lewandowski, I., Clifton-Brown, J.C., Andersson, B., Basch, G., Christian, D.G., Jorgensen, U., Jones, M.B., Riche, A.B., Schwarz, K.U., Tayebi, K., and Teixeira, F., 2003, Environment and harvest time affects the combustion qualities of Miscanthus genotype, *Agron. J.* **95**:1274–1280.

Liebig, M.A., Johnson, H.A., Hanson, J.D., and Frank, A.B., 2005, Soil carbon under switchgrass stands and cultivated cropland, *Biomass Bioenerg.* **28**:347–354.

Lynd, L.R., Haiming, J., Michels, J.G., Wyman, C.E., and Dale, B., 2003, *Bioenergy: Background, Potential, and Policy*, Policy briefing, Center for Strategic and International Studies, Washington, DC, 39 pp., http://www.csis.org/tech/biotech

Mann, C.C. and Plummer, M.L., 2002, Forest biotech edges out of the lab. *Science* **295**:1626–1629.

McCarl, B.A. and Schneider, U.A., ca. 2001, *Economic Potential of Biomass Fuels for Greenhouse Gas Emission Mitigation*, Texas A&M University, Department of Agricultural Economics, Mimeo report, unpaginated.

McCarl, B.A., Gillig, D., Lee, H-C., El-Halwagi, M., Qin, X., and Cornforth, G., 2005, Potential for bio-based greenhouse gas emission mitigation: rationale and potential, in Outlaw, J., Collins, K., and Duffield, J.A. (eds.), *Agriculture as a Producer and Consumer of Energy*, CBAI Publishing, Cambridge, Massachusetts, pp. 300–316.

McCarl, B.A., Kim, M-K., Lee, H-C., Murray, B.C., Sands, R.D., and Schneider, U.A., 2006, Insight from agricultural and forestry GHG offset studies that might influence IAM modeling, in Schlesinger, M., Kheshgi, H., Smith, J., de la Chesnaye, F., Reilly, J., Wilson, T., and Kolstad, C. (eds.), *Human Induced Climate Change*: An Interdisciplinary Assessment. Cambridge University Press, Cambridge, UK (in press). 2006.

McLaughlin, S., Bouton, J., Bransby, D., Conger, B., Ocumpaugh, W., Parrish, D., Taliaferro, C., Vogel, K., and Wullschleger, S., 1999, Developing switchgrass as a bioenergy crop, in Janick, J. (ed.), *Perspectives on New Crops and New Uses*, ASHS Press, Alexandria, Virginia, pp. 282–299.

McLaughlin, S.B. and Kszos, L.A., 2005, Development of switchgrass (*Panicum virgatum*) as a bioenergy feedstock in the United States, *Biomass Bioenerg*. **28**:515–535.

Metting, F.B., Smith, J.L., and Amthor, J.S.,1999, Science needs and new technologies for soil carbon sequestration, in Rosenberg, N.J., Izaurralde, R.C., and Malone, E.L. (eds.), *Carbon Sequestration in Soils: Science, Monitoring and Beyond: Proceedings of the St. Michaels Workshop*, December 1998, Battelle Press, Columbus, Ohio, pp. 1–34.

Metting, F.B., Scott, M.J., Benemann, J., Greenbaum, E., Seibert, M., Spormann, A., Yukawa, H., and Houghton, J., 2004, Microbial and enzyme biotechnology, in Rosenberg, N.J., Izaurralde, R.C., and Metting, F.B. (eds.), *Applications of Biotechnology to Mitigation of Greenhouse Warming: Proceedings of the St. Michaels II Workshop*, April 2003, Battelle Press, Columbus, Ohio, pp. 115–145.

Mulkey, V.R., Owens, V.N., and Lee, D.K., 2006, Management of switchgrass-dominated conservation reserve program lands for biomass production in South Dakota, *Crop Sci*. **46**:712–720.

O'Callaghan, M., Glare, T.R., Burgess, E.P.J., and Malone, L.A., 2005, Effects of plants genetically modified for insect resistance on nontarget organisms, *Ann. Rev. Entomol*. **50**:271–292.

Pacala, S. and Socolow, R., 2004, Stabilization wedges: solving the climate problem for the next 50 years with current technologies, *Science* **305**:968–972.

Parton, W.J., Stewart, J.W.B., and Cole, C.V., 1988, Dynamics of C, N, P and S in grassland soils: a model, *Biogeochemistry* **5**:109–131.

Paustian, K., Cole, C.V., Sauerbeck, D., and Sampson, N., 1998, CO_2 mitigation by agriculture: an overview, *Clim. Change* **40**:135–162.

Perlack, R.D., Wright, L.L., Turhollow, A.F., Graham, R.L., Stokes, B.J., and Erbach, D.C., 2005, *Biomass as a Feedstock for a Bioenergy and Bioproducts Industry: The Technical Feasibility of a Billion Ton Annual Supply*, Report No. DOE/GL-102005-2135, Oak Ridge National Laboratory/TM-2005/66, 59 pp.

Perseley, J.J. and Siedow, J.N., 1999, *Applications of Biotechnology to Crops: Benefits and Risks*, Council for Agricultural Science and Technology (CAST), Issue Paper No. 12. Ames, Iowa, 14 pp.

Post, W.M., Izaurralde, R.C., Jastrow, J.D., McCarl, B.A., Amonette, J.E., Bailey, V.L., Jardine, P.M., West, T.O., and Zhou, J., 2004, Enhancement of carbon sequestration in US soils, *BioScience* **54**:895–908.

Raghu, S., Anderson, R.C., Daehler, C.C., Davis, A.S., Wiedenmann, R.N., Simberloff, D., and Mack, R.N., 2006, Adding biofuels to the invasive species fire, *Science* **313**:1742.

Rosenberg, N.J., 1982, The increasing CO_2 concentration in the atmosphere and its implication on agricultural productivity. II. Effects through CO_2-induced climatic change, *Clim. Change* **4**:239–254.

Rosenberg, N.J., Izaurralde, R.C., and Malone, E.L. (eds.), 1999, *Proceedings of the St. Michaels Workshop*, December 1998, Battelle Press, Columbus, Ohio, 199 pp.

Schneider, U.A. and McCarl, B.A., 2003, Economic potential of biomass based fuels for greenhouse gas emission mitigation, *Environ. Resour. Econ.* **24**:291–312.

Shinnar, R. and Citro, F., 2006, Energy Enhanced: A roadmap to U.S. decarbonization, *Science* **313:**1243–1244.

Smith, S.J., Brenkert, A., and Edmonds, J.A., 2006, *Biomass with Carbon Dioxide Capture and Storage in a Carbon Constrained World*, Report of the Joint Global Change Research Institute, College Park, Maryland, 6 pp., seen in manuscript.

Toenniessen, G.H., 2000, *Vitamin A Deficiency and Golden Rice: The Role of the Rockefeller Foundation*, http://208.240.92.21/99prog/food/vitamina.html

Tuskan, G.A., Wullschleger, S.D., Cushman, J.H., Graham, R.L., and Thomas, S.R., 2004, Mitigation of greenhouse warming, biomass-based energy supply systems and accelerated domestication of energy crops, in Rosenberg, N.J., Izaurralde, R.C., and Metting, F.B. (eds.), *Applications of Biotechnology to Mitigation of Greenhouse Warming: Proceedings of the St. Michaels II Workshop*, April 2003, Battelle Press, Columbus, Ohio, pp. 23–59.

United Nations, 1992, *United Nations Framework Convention on Climate Change*, United Nations, New York.

United Nations, 1998, *The Convention and Kyoto Protocol, United Nations Framework Convention on Climate Change*, Bonn, Germany, http://unfccc.de/resource/convkp.html

US Congress, 2005, *Energy Policy Act of 2005: HR 6, 109*th *Congress of the United States of America*, Sections 201–241, 941–948.

US Department of Energy/Energy Information Administration, 2006, Annual energy outlook, DOE/EIA-0383 (2006), pp. 81, 84.

Verma, S.B., Dobermann, A., Cassman, K.G., Walters, D.T., Knops, J.M., Arkebauer, T.J., Suyker, A.E., Burba, G.G., Amos, B., Yang, H., Ginting, D., Hubbard, K.G., Gitelson, A.A., and Walter-Shea, E.A., 2005, Annual carbon dioxide exchange in irrigated and rainfed maize-based agroecosystems, *Agric. For. Meteorol.* **131**:77–96.

Vogel, K.P. and Jung, H.G., 2001, Genetic modification of switchgrass and other herbaceous plants for feed and fuel, *Crit. Rev. Plant Sci.* **20**:15–49.

Vogel, K.P. and Masters, R.A., 1998, Developing switchgrass into a biomass fuel crop for the Midwestern USA. Paper presented at *Bioenergy '98: Expanding Bioenergy Partnerships*, Madison, Wisconsin, October 4–8, 1998. http://bioenergy.ornl.gov/papers/bioen98/vogel.html

Vogel, K.P., Dewald, H.J., Gorz, H.J., and Haskins, F.A., 1985, *Development of Switchgrass, Indiangrass, and Eastern Gamagrass: Current Status and Future. Range Improvement in Western North America: Proceedings of the Meeting of Range Management Society*, Salt Lake City, Utah, February 14, 1985, pp. 51–62.

Vogel, K.P., Brejda, J.J., Walters, D.T., and Buxton, D.R., 2002, Switchgrass biomass production in the Midwest USA, *Agron. J.* **94**:413–420.

Waggoner, P.E., 1994, *How Much Land can Ten Billion People Spare for Nature?* Report 121, Council for Agricultural Science and Technology (CAST), Ames, Iowa, 64 pp.

Watterson, I.G., Dix, M.R., Gordon, H.B., and McGregor, J.L., 1995, The CSIRO nine-level atmospheric general circulation model and its equilibrium present and doubled CO_2 climates. *Aust. Meteorol. Mag.* **44**:111–125.

West, T.O. and Marland, G., 2002, A synthesis of carbon sequestration, carbon emissions, and net carbon faux in agriculture: comparing tillage practices in the United States, *Agric. Ecosyst. Environ.* **1**:217–232.

Wigley, T.M.L., Richels, R., and Edmonds, J.A., 1996, Economic and environmental choices in stabilization of atmospheric CO_2 concentrations, *Nature* **379**:240–243.

Wright, L.L. and Tuskan, G.A., 1997, *Strategy, Results and Directions for Woody Crop Research Funded by the US Department of Energy*, Technical Association of the Pulp and Paper Industry (TAPPI), Proceedings of the 1997 Pulping Conference, pp. 791–799.

CHAPTER 7

OUTLOOK

1. RECAPITULATION

To recapitulate the salient points made in the preceding chapters:

The NAGP, extending from North Texas into the southern portions of the Canadian Prairie Provinces, was a vast grassland before the European settlement. Tall grasses dominated the subhumid eastern region and graded with decreasing annual rainfall through a zone of intermediate grasses in the central portion to short bunch grasses in the drier west. Since settlement, the grasslands of the NAGP region have been largely converted to agriculture and ranching.

The region, covering some 2.25 million square kilometers, is one of abundant natural resources. It is endowed with mostly productive or potentially productive soils. Deep, fertile, mollisols dominate the region. All of the NAGP's soils are subject to wind and water erosion. The water resources of the region are substantial although not uniformly distributed; in some portions they are insufficient to meet all needs. The major river systems of the region flow eastward to the Mississippi and northward to Hudson Bay. The region has one major underground water resource as well—the overused Ogallala or High Plains aquifer. The water resources of the region are threatened by groundwater mining, point and nonpoint source pollution of groundwater and of runoff water to streams. The water resources could be very sensitive to the effects of climatic change.

Climates of the Great Plains range from subhumid in the east to semiarid in the west and from near dry tropical in the south to near Boreal in the north. It is the region's rapidly fluctuating atmospheric conditions (its weather) that most clearly determine the extent to which the region's agricultural and grazing potential is reached in any given year. The stressors of extreme temperatures (both high and low), shortage or overabundance of precipitation, severe and sometimes damaging winds and hail storms directly affect crop and pasture growth. These also determine the nature and severity of pests—insects, diseases, weeds—setting an upper limit on the region's agricultural and grazing potential. The climate has some good features: ample sunshine and generally dry weather at harvest offer some distinct advantages over other agricultural regions.

In spite of its climate (most particularly its frequent and severe droughts) the NAGP is one of the world's prime agricultural regions. Fifty percent of US

Great Plains land is in range, 25% in crops, 15% is in forest, with the remainder in a variety of special uses (Chapter 2). Fifty-five percent of the Canadian Great Plains land is cropped. The region produces 21% of the US corn crop and 9% of Canada's; 52% of the US wheat crop and 89% of Canada's; and 93% of US canola crop and 98% of Canada's. Together the US and Canadian Great Plains supply 8% of the world's corn, 8% of its wheat, and 12% of its canola. The region also produces 50% of US cattle and 50% of Canada's; 12% of US hogs and 25% of Canada's; and 37% of US sheep and lambs and 30% of Canada's.

These impressive statistics notwithstanding, all is not well on the Plains. Once thought to be a "Great American Desert," the region seemed to its earliest explorers an unlikely candidate for agrarian settlement. Yet, even before the Civil War, settlers were pushing at its eastern edges. Ever since the process began, the wisdom of breaking the sod and removing the original grass cover to make way for agricultural crops has been vehemently debated. Imperfect adaptation to its natural limitations, inappropriate settlement policies, management failures, and what some consider perverse economic incentives raise questions yet today about the region's long-term sustainability. Since settlement began there has been conflict over the proper land use for the region. Boosters (whatever their motives) encouraged farming on the Plains. Government policies such as the Homestead Act favored it. Still today, government price supports and other forms of payment continue to encourage farming on the Plains.

Surveyors like Powell in 1879, grounded in the science of that time, argued for maintaining grass cover on as much of the land as possible. So, too, did the *Report on the Future of the Great Plains* presented in 1936, during the dust bowl era, to President Franklin D. Roosevelt. Today, the notion of a "Buffalo Commons", a return of much of the western Plains to grass with buffalo herds roaming in it, has its adherents. The sustainable agricultural movement, while somewhat more nuanced in its approach, also strongly favors a return of much of the land to native grasses, or at least better managed grasslands for grazing.

Whatever the merits of these arguments, there are some distressing facts about the Great Plains—at least its still rural portions. The region is poor and highly dependent in the USA upon government payments for survival. The population is declining in many but not all rural counties, and is certainly declining as a proportion of the total population. Although US and Canadian government programs since the 1930s have provided significant help, the region remains extremely vulnerable to the effects of drought, whether protracted as in the 1950s or short and intense as in 1976–1977. As this is written (August 2006) much of the US portion of the Plains (all of Montana, Wyoming, and the western halves of North Dakota, South Dakota, and Nebraska) was again experiencing severe drought. Portions of the Prairie Provinces were also experiencing significant shortages of rainfall at that time, especially southeastern Manitoba.

Against this backdrop comes an additional complication—the prospect of climatic change. Virtually all scientists who have studied the matter are convinced that the continuing emissions of greenhouse gases into the earth's atmosphere will lead to global warming and a consequent change in climate worldwide. Indeed the evidence strongly suggests that the warming is already occurring. Although the timing and geographic distribution of "greenhouse-forced" climatic change are still quite uncertain, it seems more probable than not that temperatures will rise in the Great Plains region—probably more so in the north than in the south. The amounts, intensity and timing of precipitation are also likely to change. Because so much of the region is "on-the-edge" in terms of extreme temperatures and marginal precipitation, it seems likely that the effects of climate change will be significant and more profound than it might be in other adjacent and more amply watered or more temperate climatic zones such as the Cornbelt. It is important to reemphasize here that while how, specifically, the Plains region will be affected by climatic change is uncertain, global warming-forced climatic change, *per se,* is virtually certain in the region's future.

Simple enough to say that since crops respond to temperature their yields will be affected by climate change. Simulation studies using the most credible of the general circulation model projections of future climatic changes suggest reduced productivity for the Plains. But rising temperatures can bring positive changes such as longer growing seasons for annual crops in cool, high latitude zones. The most credible projections suggest reductions in precipitation. Changes in precipitation regime will directly affect crop and rangeland productivity as well as water supplies for irrigation, navigation, recreation, municipal and industrial uses. Certain of the direct effects of changing temperature, precipitation, humidity, etc. could be offset to a degree by the rising atmospheric concentration of CO_2 which tends to stimulate photosynthesis while reducing plant water consumption (transpiration). But, as explained in the foregoing chapters, there are no sure bets in the lottery of climate change, and while potentially positive impacts are possible, greater rather than lesser climatic stresses are more likely in the future. It would be unwise for farmers and policymakers anywhere to bank on potentially positive climatic changes.

The aforementioned impacts (which do not consider farmer adaptations) could be quite serious. However, there is some reason to believe that, given the necessary budgetary support and absent extreme changes in climatic conditions, agricultural science, which has demonstrated remarkable adaptability and versatility in the face of drought and other stressors (recall Figures 6-7 to 6-9) will provide tools needed to keep losses to a minimum. Logically, the science establishment should already be at work developing adaptations to climatic change by, for example, breeding cultivars with greater tolerance to drought and high temperatures. Adaptations of this kind make sense as "no-regret" strategies. Even if climate change fails to materialize, they are needed to better cope with the current climate of the Great Plains and of similar regions. But adaptation can carry us just so far. It is no less important that a serious strategy for the mitigation of greenhouse warming be developed and implemented. This is a job for all governments and all peoples, Plainsmen and women included. And of primary importance if greenhouse warming is to be moderated or avoided is the reduction in greenhouse gas emissions, primarily through reduced consumption of fossil fuels.

2. BIOMASS AND THE GREAT PLAINS

The NAGP is not the only agricultural region that can contribute to the goal of mitigating climate change, but it can contribute importantly and may have more to gain in both absolute and relative terms than other regions. One opportunity is through restoration of carbon to its soils which, by increasing soil fertility and moisture-holding capacity and protecting against wind and water erosion, has the potential of improving and stabilizing crop yields in the region. Another opportunity is through the production of biomass which can be used to reduce the need for fossil fuels and provide important new economic opportunities for the region's farmers and for agricultural industry.

The first of these climate change mitigation methods—restoration of soil carbon offers a relatively simple and inexpensive way of combating climate change. As explained in foregoing chapters, the soils of the Plains lost a large portion of their carbon-containing organic matter when they were converted from grassland to farming. The carbon in that organic matter was oxidized to form CO_2 which diffused into the atmosphere. Some of the lost carbon can be returned to the soil and sequestered there. Additionally, the farm management practices that foster soil carbon sequestration, such as minimum and zero tillage, reduce the energy requirements in agriculture and are environmentally benign in other ways.

The second of the mitigation methods—production of biomass for direct firing and to produce transportation fuels—offers a practical way to significantly reduce both the need for fossil fuels and CO_2 emissions to the atmosphere. The opportunities that biomass provides and the problems it raises are the major focus of this book. Although they have been discussed in considerable detail in the foregoing pages, a few key issues with regard to biomass production on the Plains should be reemphasized here.

A substantial research effort has identified switchgrass as well suited for biomass production in the US Midwest and South and in eastern Canada. Its suitability for the eastern Plains has also been demonstrated. Because of its warm season metabolism and deep and extensive root system, this species is better suited than the traditional crops grown on the Plains to the high temperatures and moisture stress of projected climatic change for the region. Of course, corn and soybeans, the crops that are currently most important in production of ethanol and biodiesel, also grow well in the eastern Plains. And the ample supplies of corn stover and other crop residues to be found there can be used as feedstock for cellulosic ethanol production.

Less effort has been devoted to identification and management of plant species that can produce biomass reliably in the drier western portions of the Plains, although research is now intensifying there. It is the agriculture of this drier region whose ecological and economic sustainability is most precarious. Biomass production may offer an opportunity to effect land use changes that foster sustainability in that portion of the Plains and provide new economic opportunities for its farmers and wider employment opportunities for its citizens.

Legitimate concern has been raised about whether conversion of large areas of US and Canadian crop producing land to biomass production will diminish food and feed supplies and lead to shortages and price increases that ultimately upset domestic consumers and international markets. Some of this concern can be allayed by the recognition of the fact that as the result of a steady stream of improved varieties, fertilization, and other management practices, crop and forage yields continue to increase steadily in the USA and Canada, as they have since the mid-20th century. Traditional plant breeding augmented by genetic engineering should help continue, if not accelerate, this trend and, in so doing, help offset some of the effects of land conversion from food and feed crops to biomass. Genetic engineering can also contribute to improvement in quality of food and biomass crops; in the latter case by, for example, increasing cellulose content to enhance ethanol yields and/or by suppressing flowering to reduce the risks of undesirable gene transfer to other related plant species.

Converting large portions of the Plains to production of biomass crops will not satisfy the ideal of a return to pristine grassland or even, for that matter, to a tourist-attracting "Buffalo Commons". But since perennial grasses provide the soil cover that protects against wind and water erosion, provides habitat for wildlife and could provide emergency forage in times of drought, it would surely be a move in an ecologically sound direction. Once established, perennial grasses are environmentally benign in that they require less fertilizer and chemical pest control than do the annual food and feed crops they would replace. And after stand establishment the grasses require no tillage for 8 or 10 years until the biomass crop is rotated into other crops or a new stand of grasses is planted.

3. IS THERE A BIOMASS FUTURE?

In view of all of the foregoing information, what are the general prospects for a biomass-based industry? Four trends converging at this time appear to favor its prospects. These can be described under four headings: environmentalism, "petropolitics", rising energy prices, and achievements in biomass-related research.

Environmentalism: The first trend to note is growing public concern for "the environment" which manifests itself in many ways. Most relevant here is what this book calls "The Wildcard of Climate Change". We are virtually certain that the world will continue to warm with continuing emissions of greenhouse gases into the atmosphere and that climate will change throughout the world although we are much less certain about how and where these changes will be manifested. Despite the remaining uncertainties, public concern about global warming has been rising steadily in the USA, Canada, the European Union, and elsewhere.

The need to reduce CO_2 emissions is understood by the public generally, not only by climate change specialists and environmentalists. Since the combustion of coal, oil, and natural gas is the principal source of the CO_2 emissions, demand is growing for technologies that enable renewable energy resources to be used as substitutes for these fuels. That biomass appears to be among the least environmentally risky alternatives for reducing current use of fossil fuels increases public interest and support for it.

One of the important obstacles to wider acceptance of biomass has been the concern that the net energy balance (NEB) of its most prominent product thus far, ethanol from corn, is modest and that its price has been artificially lowered by various forms of government support. Recent research indicates that the NEB of corn-based ethanol—about 25%—is modest, but that of biodiesel from soybeans is more than 90%.The NEB of cellulosic ethanol is expected to be better than that of ethanol from corn.

The surging interest in prospects for biomass for ethanol and other purposes is evidenced in the media. In an article entitled "The Race Against Climate Change", in the *Business Week* of December 12, 2005 reports that leaders of greenhouse gas emitting industries, anticipating mandatory limits on emissions, are already moving to measure and slash their greenhouse gas emissions: "One new twist in the discussion of global warming is the arrival of a corps of sharp-penciled financiers. Bankers, insurers and institutional investors have begun to tally the trillions of dollars in financial risks that climate change poses." Major corporations have announced plans to develop activities contributing to the mitigation of climate change. General Electric, for example, has promised to double its investments in environmental research and to lower its emissions modestly.[1] Another good example of this trend is the environmental policy announced in 2006 by the investment banking firm Goldman Sachs. That policy states: ". . . we will work to ensure that our people, capital and ideas are used to help find effective market-based solutions to address climate change and other critical environmental issues, and we will seek to create new business opportunities that benefit the environment."[2]

Further evidence of the growing power of public concern with climate change, if such is needed, appears in US President George W. Bush's call in his 2006 "State of the Union" address to Congress[3] for research and development leading to methods of producing ethanol from crop and woody waste materials (cellulosic ethanol, in other words). "Our goal is to make this new kind of ethanol practical and competitive within six years", he stated. This from the

[1] http://www.businessweek.com/magazine/content/05_50/63963401.htm Find G.E. announcement on the web

[2] http://www2.goldmansachs.com/our_firm/our_culture/corporate_citizenship/environmental_policy_framework/index.html

[3] http://www.cnn.com/2006/POLITICS/01/31/sotu.transcript/

head of an administration apparently unconvinced of the reality and/or potentially serious consequences of global warming and staunchly opposed to the imposition of any but voluntary greenhouse gas emissions controls!

Petro-politics: The second trend is geopolitical in nature. Nations whose economies depend on imported petroleum have come to appreciate their sensitivity to global "petro-politics" and vulnerability to the consequences should global terrorism be focused on petroleum exporting countries. As so large a share of the world petroleum market is supplied by nations that are currently either hostile to USA or western interests (e.g., Iran, Venezuela) or vulnerable to takeover by groups hostile to the west (e.g., Saudi Arabia, Iraq, the Gulf States, Chad, Nigeria, Libya), the world has become increasingly vulnerable to "petro-blackmail". For decades lip service has been paid by politicians in energy importing nations to the notion of "energy independence." Global political instability of recent years has made energy independence a much more urgent matter than it has been heretofore in the importing countries. Biomass-based transportation fuels can contribute to that goal and reduce our reliance on petroleum and the volatile and potentially unstable petroleum-exporting nations. Brazil, which is now virtually independent of foreign oil imports because of its national ethanol production capacity has become a model that other countries would like to copy. Over 18.2 billion liters (4.8 billion US gallons) of ethanol were produced from sugarcane in Brazil in 2005. Of this, more than 2.27 billion liters (0.6 billion gallons) were exported.[4]

Energy prices: Another trend or factor not entirely dissociated from petro-politics that has raised interest in biomass in public, policymaker, industrial and entrepreneurial circles is the sharp rise in the market price of energy of all forms during the middle years of this decade. This rise can be attributed in part to political instability in some of the petroleum exporting countries. It can also be attributed, simply enough, to demand outpacing supply, a situation prompted by increasing global population and improving standards of living. A major factor in increasing demand and competition for energy sources has been the rapid development and emerging economic power of China and India and to a lesser extent of other rapidly developing nations. Notable, if temporary, decreases in petroleum production and processing caused by natural disasters such as Hurricane Katrina striking the US Gulf Coast in September 2005 and by engineering failures such as the leakage from British Petroleum's Alaskan pipelines in August 2006 have also contributed to the sharp spikes in petroleum prices. In the of summer of 2006 these factors, together, drove the price of regular grade gasoline to over \$3.00 per gallon (\$0.79 per liter) in the USA (with still higher prices elsewhere). This price is well above the \$1.65–1.80 per gallon range, which only a few years ago economists speculated would be required to make biomass ethanol competitive with gasoline.

---

[4] http://www.card.iastate.edu/iowa_ag_review/spring_06/article3.aspx

Achievements in biomass research: The practicality of producing biomass crops in the southern and Midwestern US has been demonstrated in long-term research efforts supported by the USDOE and USDA. Field tests of existing and improved cultivars of poplar, switchgrass, and other candidate biomass crops have demonstrated the feasibility of producing them under optimized and sustainable management. The poplar genome has now been successfully mapped (Tuskan et al. 2006) and together with that of switchgrass, when complete, will afford opportunities to genetically engineer these species (and likely others in the future) to increase their productivity, harvestability, and processing characteristics.

The energy in cellulose is locked away in sugars that resist biological and chemical degradation, so that cellulosic ethanol is more difficult to produce than ethanol from grain. Research will be required on enzymatic mechanisms for breakdown of the cellulosic biomass and on processes that reduce cost and overall production. New enzymes (some of them genetically engineered) are being developed in the laboratory and subjected to testing in pilot-scale "biorefineries". The DOE's July 2006 report "Breaking the Biological Barriers to Cellulosic Ethanol" states that recent breakthroughs in biotechnology make possible the release of energy in the sugars that constitute cellulosic materials.[5] It asserts that the replacement of 30% of US gasoline use with biofuels by 2030 is an "attainable goal". Cellulosic ethanol appears to offer the best opportunity for a major increase in the role of renewable fuels. In the long term the contribution that biomass makes to energy independence and greenhouse gas abatement in North America will depend on technological advances and economics circumstances that allow cellulosic biomass to gain a significant market share.

4. CONCLUDING THOUGHTS

The factors discussed above—environmentalism, petro-politics, the demand for energy independence, high energy prices, the putative greening of big industry, and the interest of financial markets—have given biomass a big push forward during the past few years. Biomass seems to have a global future and seems promising for already important food-producing portions of North America. And that form of "biomass" that is soil carbon sequestration is achievable and beneficial virtually anywhere that crops are grown.

Introducing dedicated biomass crops into the regular crop rotation can offer ecological and agronomic advantages in all agricultural regions. With power plants and biorefineries located accessibly, biomass can offer farmers a steadier cash flow than other, more traditional crops, especially if (when?) price supports are withdrawn from the latter. This view of the future applies as well to

[5] Breaking the biological barriers to cellulosic ethanol: a joint research agenda. A research roadmap resulting from the Biomass to Biofuels Workshop. December 7–9, 2005. Rockville, MD. DOE/SC 0095. Released June, 2006.

the eastern Plains as it does to the Cornbelt or the South. In the case of the drier portions of the Great Plains where biomass crops will likely yield less, and transportation costs to power plants and biorefineries could be greater, prospects for a biomass economy are less clear.

A strong argument for biomass cropping in the western Plains is ecological. It is well understood that the region would benefit from an increase in its land area under grass cover. That cover need not be total or permanent but can be included in long-term rotations including wheat, barley, rye, oats, other small grains, or canola. Straw from these crops can also be harvested for biomass purposes. Additionally, the US portion of the Plains has a large proportion of its land enrolled in the Conservation Reserve Program and, as explained in Chapter 6, the vegetation on enrolled lands can be harvested when needed for biomass with little loss of the protection afforded against soil erosion. Although not much research on this matter has been reported, it seems logical that fenced-off areas of range can also be rotated out of grazing and its vegetation allowed to grow until enough accumulates for periodic harvests of biomass.

The extensive irrigation infrastructure that now exists on the Great Plains is used primarily to grow corn and sorghum for grain, for cotton and, to a lesser degree, for wheat, alfalfa, and some specialty crops such as dry beans, sugar beets and potatoes. Concern for the sustainability of agriculture and water security in the region challenges the practice of mining water from underground aquifers in order to irrigate low-value crops like corn and sorghum. This concern is especially strong for aquifers that recharge only slowly, if at all. The current trend toward a reduction of irrigated area in the Plains states that would help to conserve its water resource may, perversely, be reversed by increased demand for grains to make ethanol. An alternative use of the existing irrigation infrastructure could be for limited and infrequent watering of biomass fields to supplement rainfall in seasons and years of shortage. Although not easily done, center-pivot sprinkler systems can be moved from field to field as needed. Other types of sprinkler systems are more easily moved. Biomass crops can also be grown in fields already fitted out for surface irrigation. Agronomic, engineering and economic analyses are needed to establish the practicality (or lack thereof) of these notions on what one might term "Strategic Biomass Irrigation".

In the years before the big run-up in gasoline prices, specialists suggested that transporting biomass for more than 50 miles (80 km) would be too costly. One disadvantage that biomass farming may encounter on the drier western Plains is that, either because of insufficient land committed to biomass production or because of relatively low yields, power plants and biorefineries there will need to draw materials from greater distances. New economic realities could alter the calculus and permit extension of the area from which individual processing plants draw their supplies of biomass. Can we not also envision a situation that justifies the deployment of a denser network of smaller-scale facilities, especially biomass-fired or co-fired electrical power plants that would reduce the need to transport biomass for great distances? Could densification of harvested biomass with machinery that tightly compacts grasses, stover, straw or other waste materials be used to significantly reduce the volume of material to be transported to the processing site? Can biomass headed to the refinery be subjected to partial chemical and/or mechanical processing before it leaves the farm or at local small-scale facilities to reduce the mass of material to be transported?[6] Most intriguing of all these "futuristic" notions is the possibility (discussed in Chapter 6), that the CO_2 released from biomass-burning power plants might be captured at the smokestack and sequestered in geological strata underlying the Plains region. In such a case, biomass would be a means of achieving not only near-zero CO_2 emissions but negative emissions leading to a decrease of the atmospheric concentration of this potent greenhouse gas.

We return, in closing, to the very supportable premise that the Plains region needs more grass in its lands. And its people need greater economic opportunity than the current land-use patterns afford. The construction and operating of biomass-fired power plants and biorefineries and the growing, harvesting, and transportation of biomass to these facilities will provide new employment opportunities to the region.

Of course, the Plains will remain a difficult region to farm. Droughts and other climatic hazards will not cease to stress the region because we will them to. But an agriculture that includes a significant regrassing of the Plains whether prompted by the need for biomass or for other reasons will be a better place for its inhabitants, a contributor to its own well-being as well as to national and global economic and ecological health.

REFERENCES

Tuskan, G., Di Fazio, S., et al: 2006, The genome of the Black Cotton wood, *Populus trichocarpa* (Tonn & Gray). Science 313, 1596–1604.

[6] One intriguing example of this is the suggestion in Shinnar and Citro (cited in Section 6.2.1, Chapter 6) of converting biomass to methanol in small local plants and transporting it to a biorefinery or existing petrochemical plant for further processing.

INDEX

E

F

G

COLOR PLATES

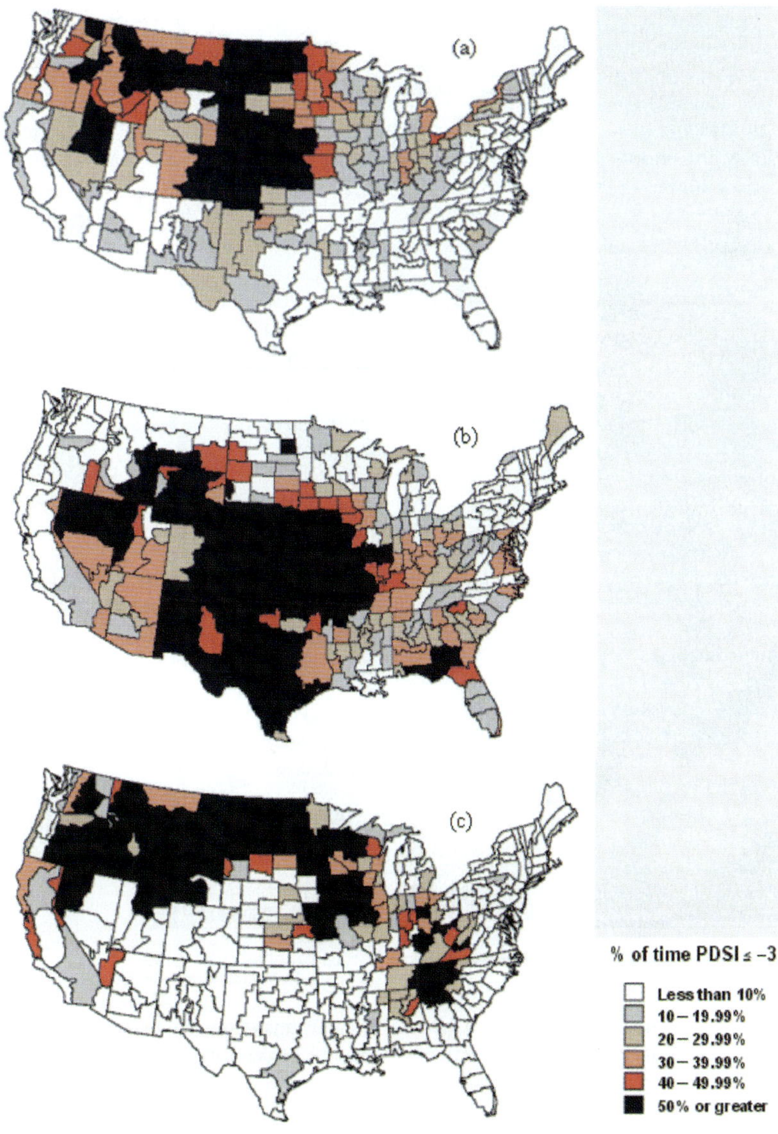

Figure 2-5. Palmer Drought Severity Indices (percent of time in severe and extreme drought) for three droughts in the USA (**a**) 1934–1939; (**b**) 1954–1956; and (**c**) 1988 (Maps prepared from various sources by the National Drought Mitigation Center, University of Nebraska, Lincoln.)

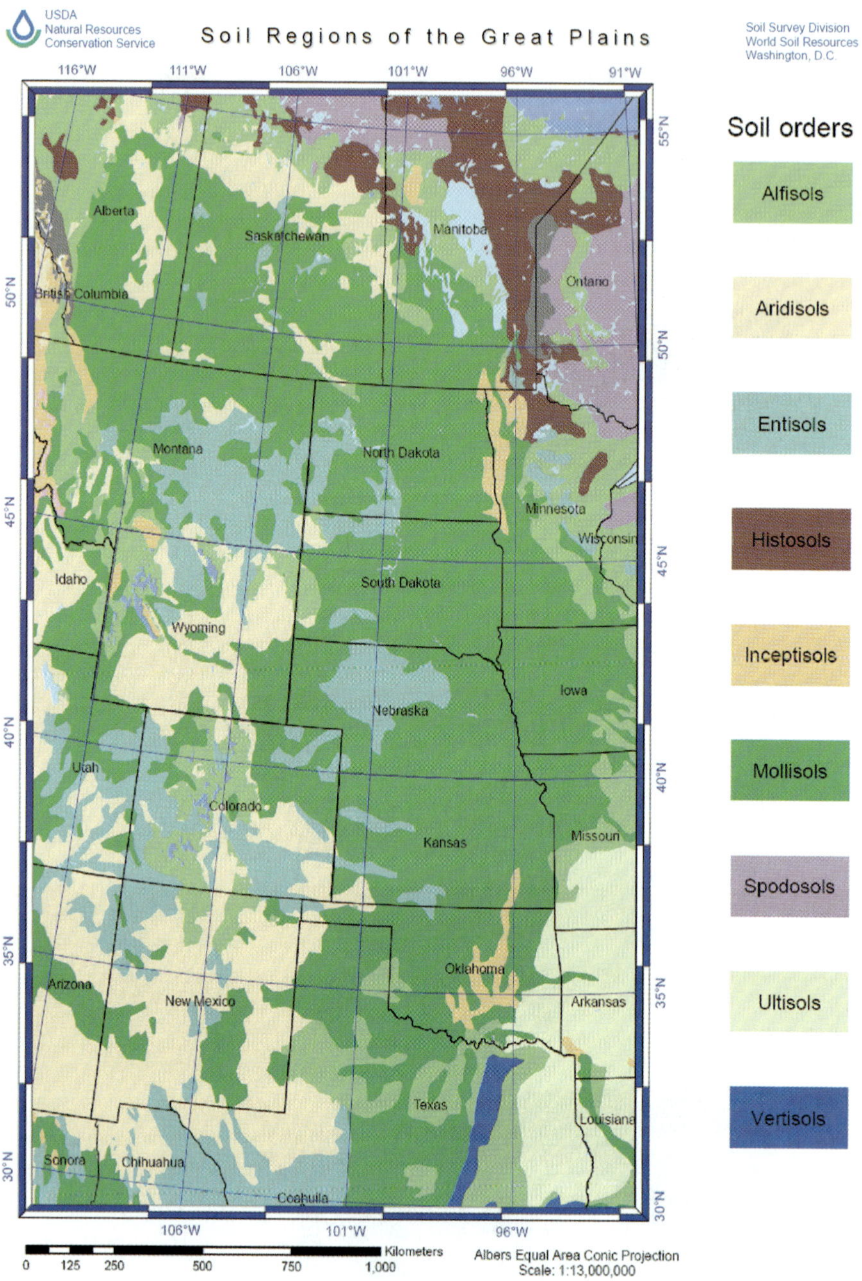

Figure 2-6. Soil regions of the North American Great Plains (Courtesy of US Department of Agriculture, Natural Resources Conservation Service, Soil Survey Division, World Soil Resources, Washington, DC.)

Figure 2-7. A typical Mollisol showing its dark colored surface horizon relatively high in content of organic matter (Adapted from US Department of Agriculture, Natural Resources Conservation Service, http://soils.usda.gov/technical/classification/orders/mollisols.html)

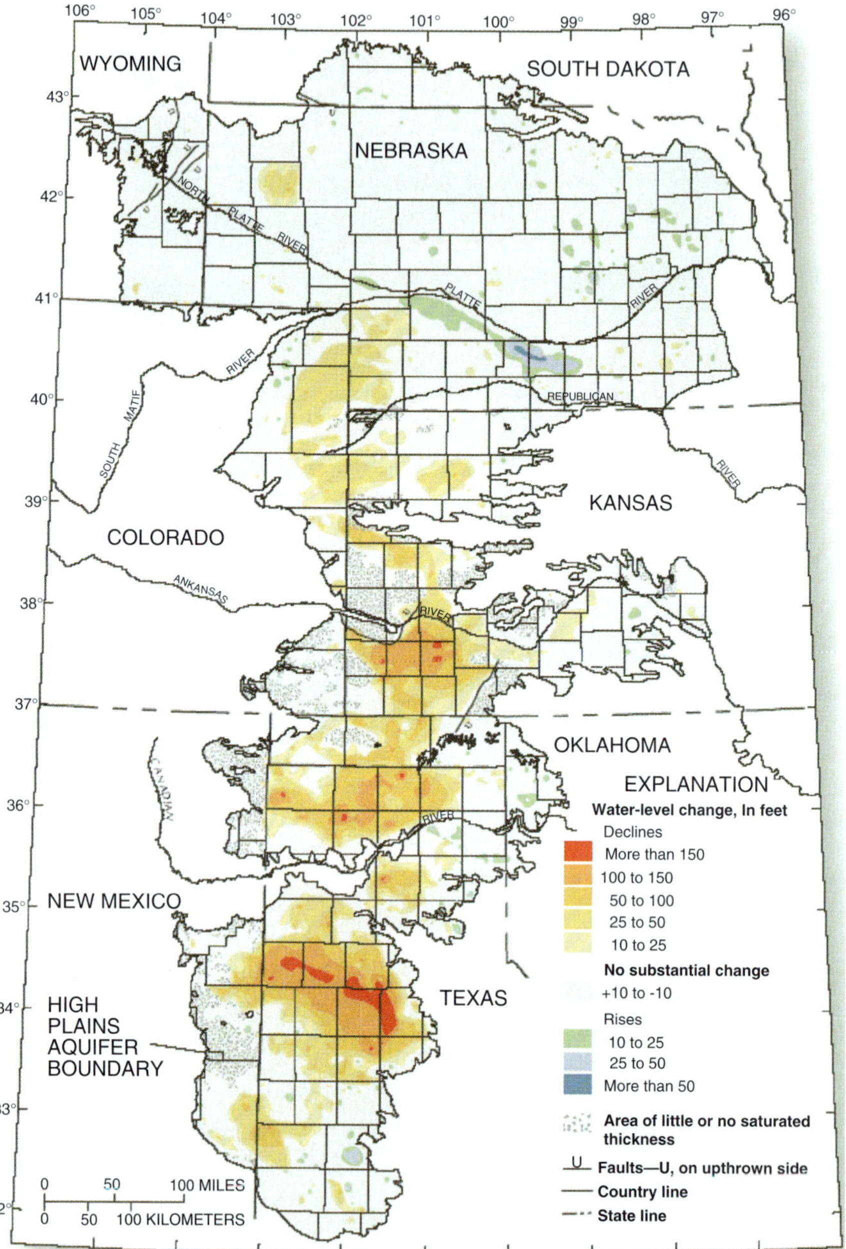

Base from U.S. Geological survey digital data, 1:2,000,000
Albers Equal-Area projection, Horizontal datum NAD83,
Standard parallels 29°30′and 45°30′, central meridian-101°

Figure 2-8. Water-level changes in the High Plains aquifer, predevelopment to 2003 (McGuire 2004, http://pubs.usgs.gov/fs/2004/3097/pdf/fs-2004-3097.pdf)

Figure 2-10. A view of shortgrass prairie near Ft. Collins, Colorado. (Adapted from Long Term Ecological Research Network, http://savanna.lternet.edu/gallery/sgs/SGS_010016_1)

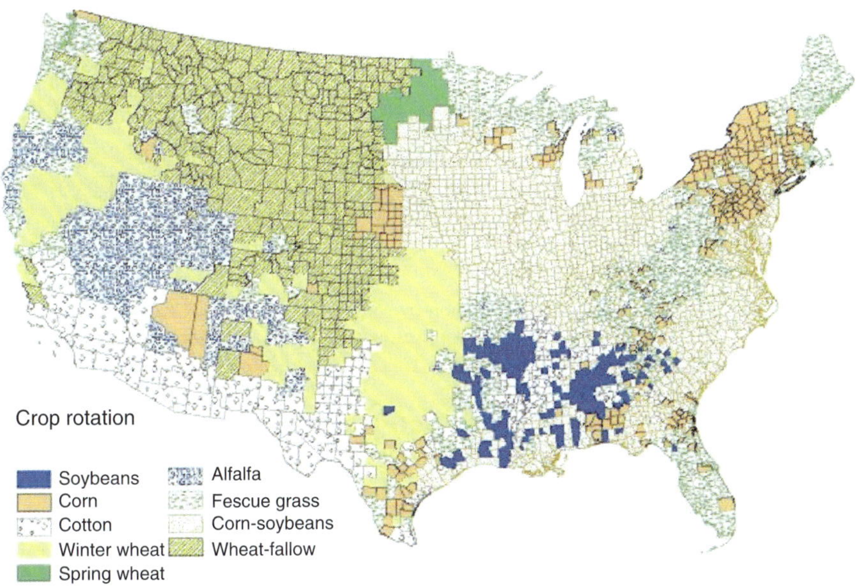

Figure 4-1. The predominant crop by county averaged from the 1985 to 1997 US Natural Resources Inventory. (Courtesy of C. Brosch and R.C. Izaurralde, Joint Global Change Research Institute, College Park, Maryland)

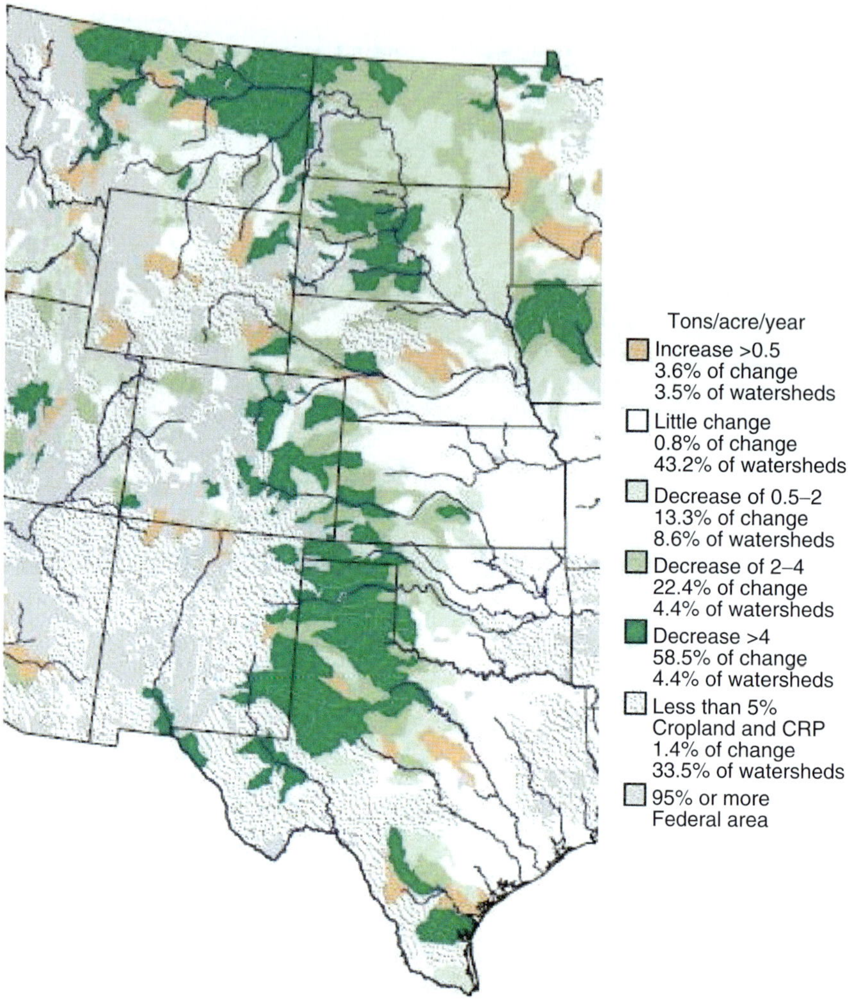

Figure 4-5. Change in average annual soil erosion by wind on cropland and CRP land, 1982–1997. (Source: US Department of Agriculture Natural Resources Conservation Service, http://www.nrcs.usda.gov/technical/land/erosion.html)

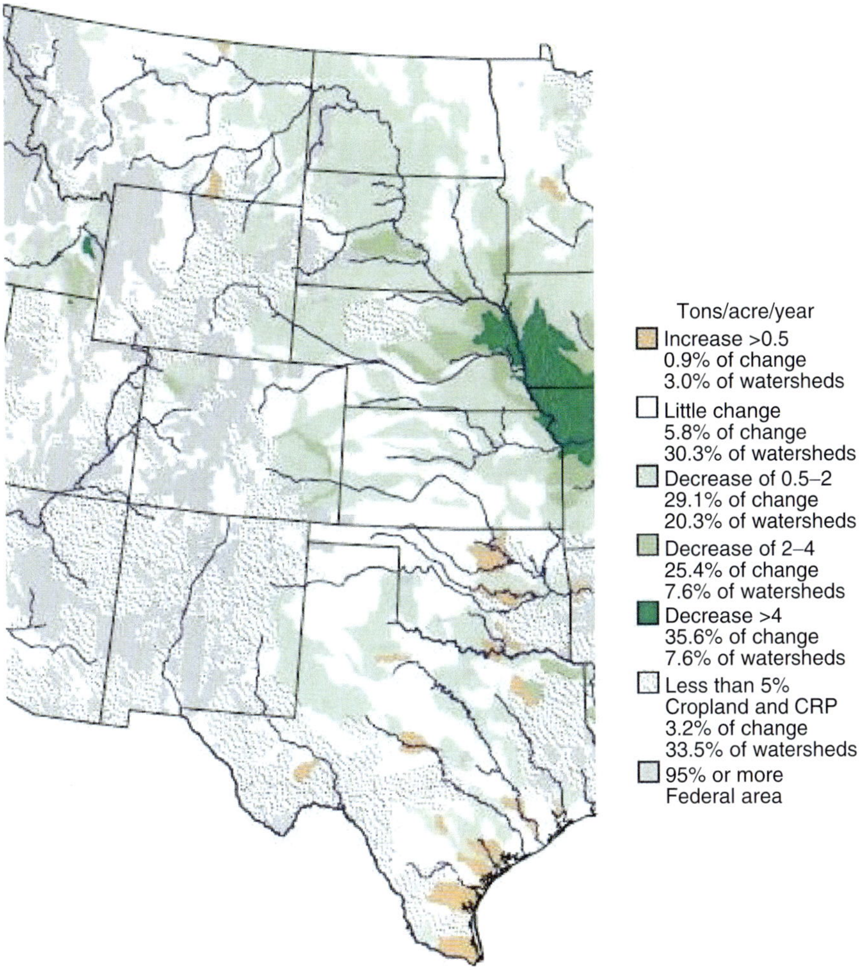

Figure 4-6. Change in average annual soil erosion by water on cropland and CRP land, 1982–1997. (Source: US Department of Agriculture Natural Resources Conservation Service, http://www.nrcs.usda.gov/technical/land/erosion.html)

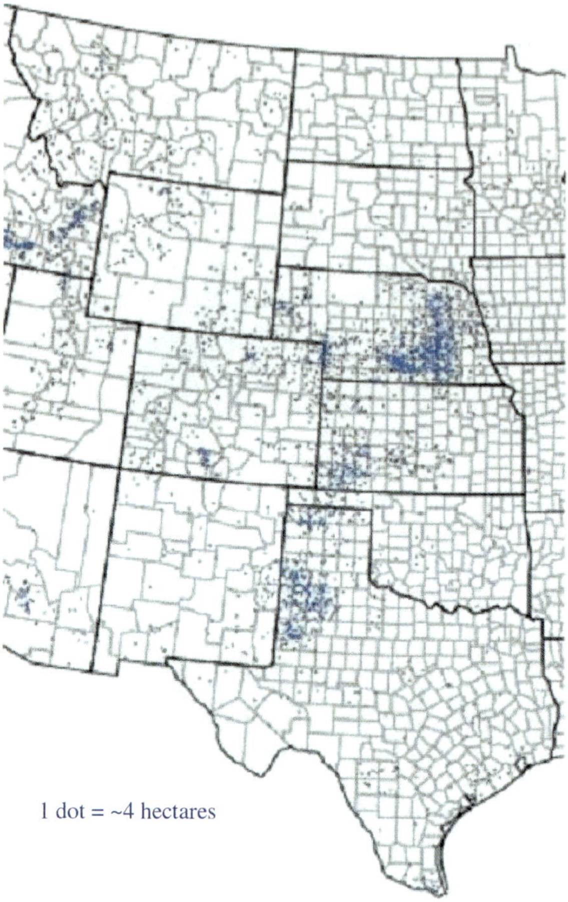

Figure 4-7. Irrigated land in the USA, 2002. (Source: 2002 US Census of Agriculture, Map 02-M079)

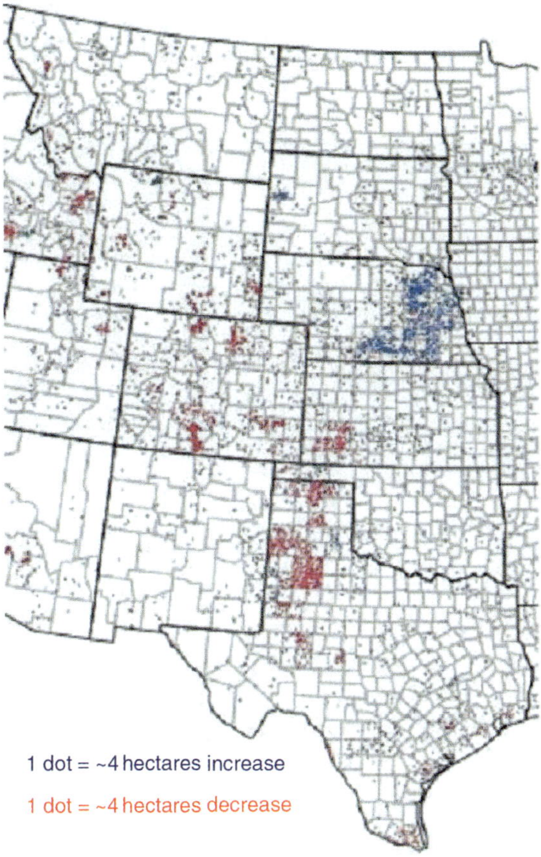

Figure 4-8. Change in area of irrigated land in the USA, 1997–2002. (Source: 2002 US Census of Agriculture, Map 02-M080)

Figure 4-9. Furrow irrigation with gated-pipe. (Source: http://www.wtamu.edu/~crobinson/Irrigation/furgateinfo.html)

Figure 4-10. Center pivot irrigation system. (Source: http://www.ars.usda.gov/is/graphics/photos/oct00/k9072-1.htm)

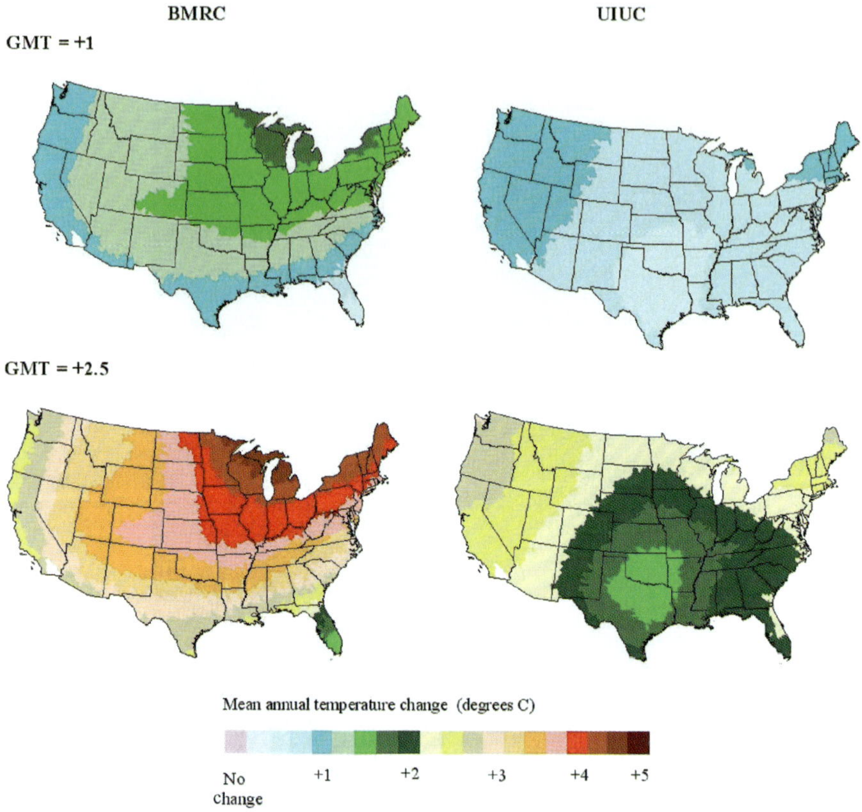

Figure 5-3a. Mean annual temperature change from baseline for the BMRC and UIUC GCMs used in the JGCRI study (Source: Smith et al. 2005)

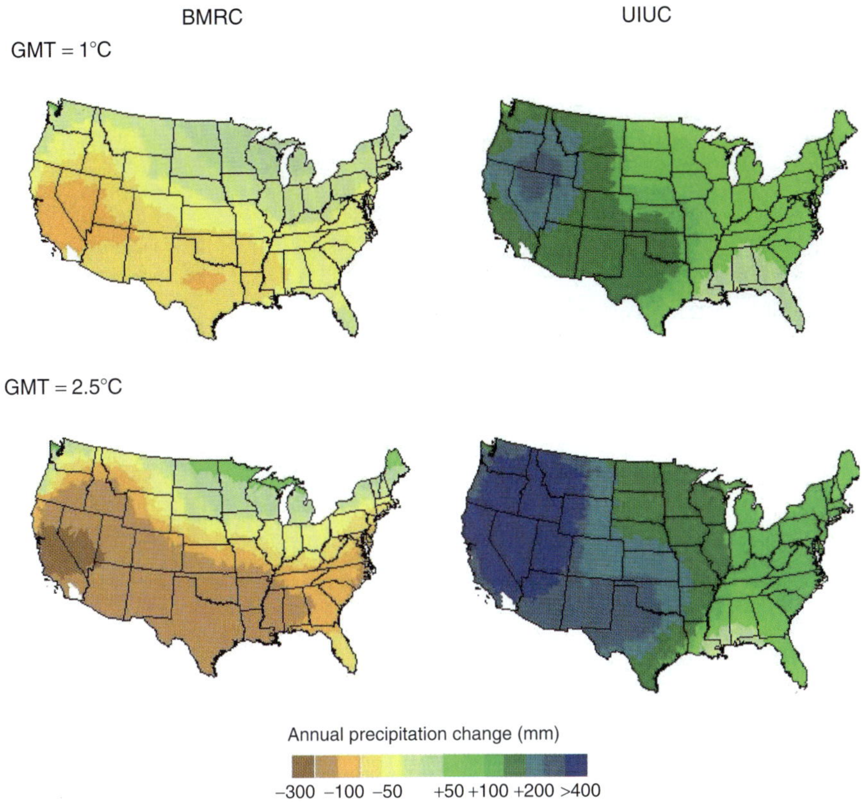

Figure 5-3b. Mean annual precipitation change from baseline for the BMRC and UIUC GCMs used in the JGCRI study (Source: Smith et al. 2005)

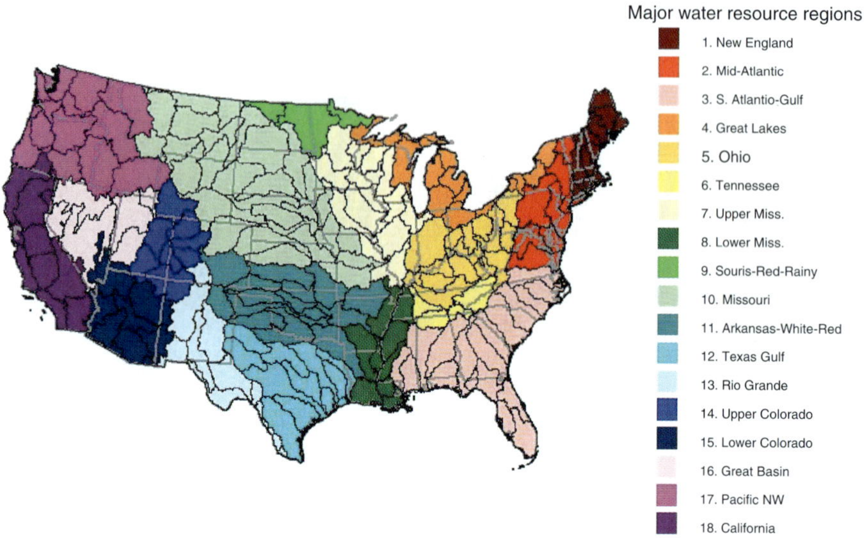

Figure 5-4. Major Water Resource Regions of the conterminous USA as defined by US Geological Survey (1987). The 204 modeling regions used in the JGCRI study are shown

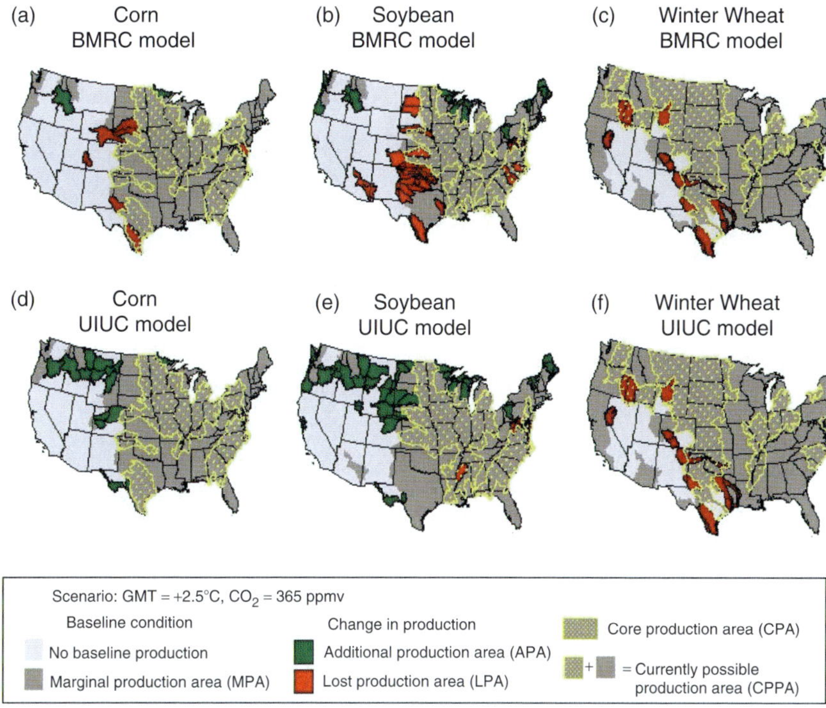

Figure 5-5. Regions projected to enter or leave production for three grain crops with the BMRC and UIUC GCMs at a global mean temperature increase of +2.5°C and CO_2 concentration of 365 ppmv (Source: Thomson et al. 2005a)

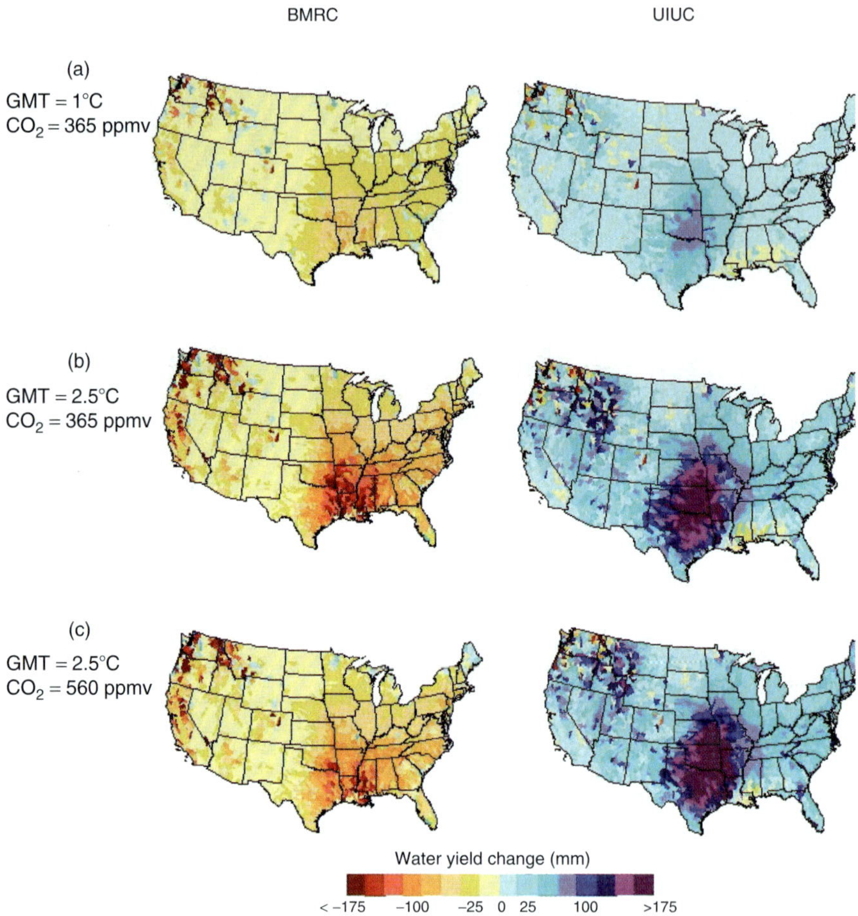

Figure 5-6. Water yield change from baseline (mm) for two GCMs with increasing global mean temperature (GMT) with and without the CO_2-fertilization effect (Thomson et al. 2005b)

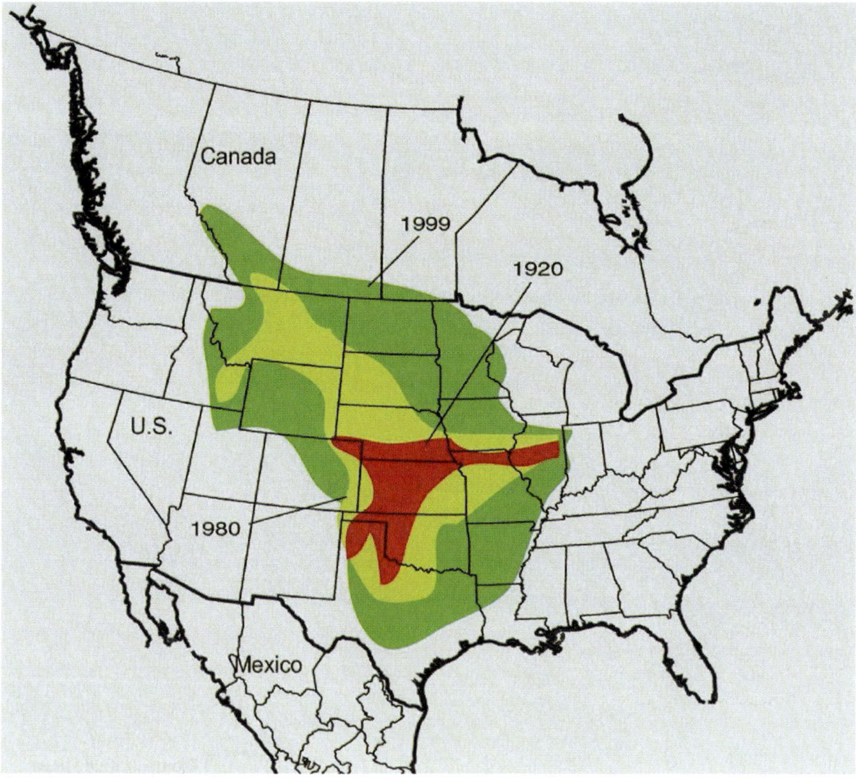

Figure 6-9. Spread of winter wheat culture from 1920 to 1999. Figure covering 1920 and 1980 from Rosenberg (1982); updated to 1999 by W.E. Easterling in 2004

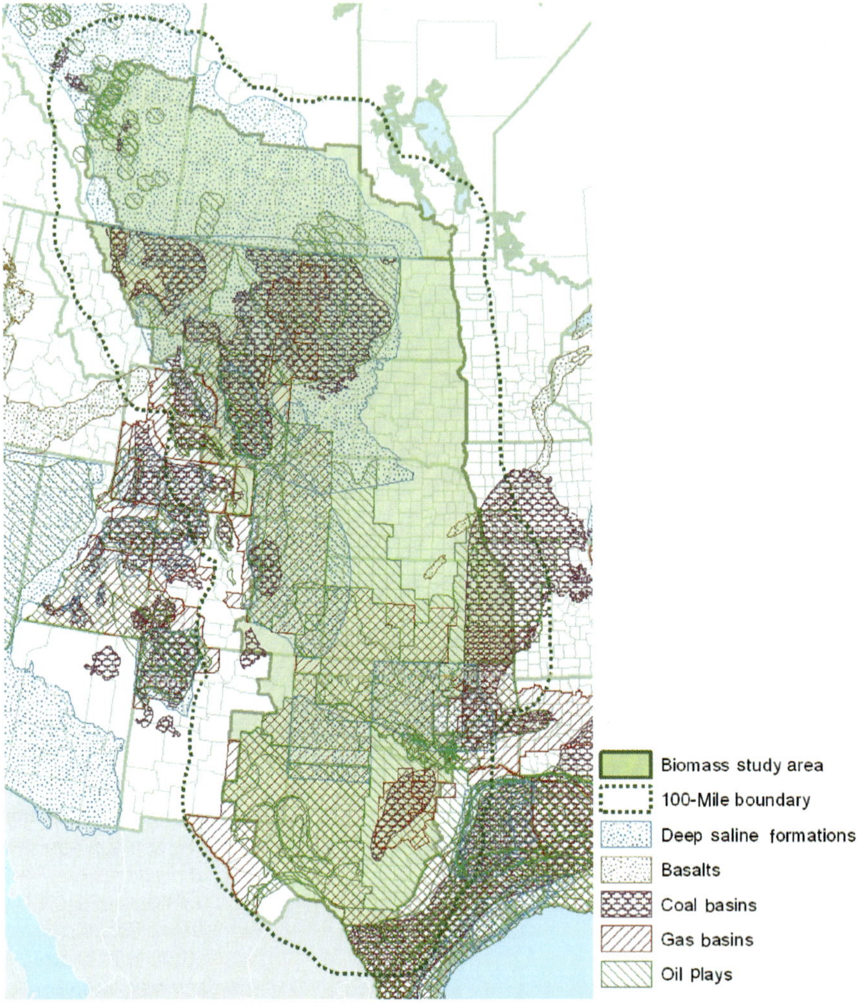

Figure 6-10. Geologic CO_2 storage potential within the North American Great Plains and the surrounding 160 km (100 miles). (Courtesy of J.J. Dooley and C.L. Davidson, Pacific Northwest National Laboratory)

Advances in Global Change Research

1. P. Martens and J. Rotmans (eds.): *Climate Change: An Integrated Perspective.* 1999 ISBN 0-7923-5996-8
2. A. Gillespie and W.C.G. Burns (eds.): *Climate Change in the South Pacific: Impacts and Responses in Australia, New Zealand, and Small Island States.* 2000 ISBN 0-7923-6077-X
3. J.L. Innes, M. Beniston and M.M. Verstraete (eds.): *Biomass Burning and Its Inter-Relationships with the Climate Systems.* 2000 ISBN 0-7923-6107-5
4. M.M. Verstraete, M. Menenti and J. Peltoniemi (eds.): *Observing Land from Space: Science, Customers and Technology.* 2000 ISBN 0-7923-6503-8
5. T. Skodvin: *Structure and Agent in the Scientific Diplomacy of Climate Change.* An Empirical Case Study of Science-Policy Interaction in the Intergovernmental Panel on Climate Change. 2000 ISBN 0-7923-6637-9
6. S. McLaren and D. Kniveton: *Linking Climate Change to Land Surface Change.* 2000 ISBN 0-7923-6638-7
7. M. Beniston and M.M. Verstraete (eds.): *Remote Sensing and Climate Modeling: Synergies and Limitations.* 2001 ISBN 0-7923-6801-0
8. E. Jochem, J. Sathaye and D. Bouille (eds.): *Society, Behaviour, and Climate Change Mitigation.* 2000 ISBN 0-7923-6802-9
9. G. Visconti, M. Beniston, E.D. Iannorelli and D. Barba (eds.): *Global Change and Protected Areas.* 2001 ISBN 0-7923-6818-1
10. M. Beniston (ed.): *Climatic Change: Implications for the Hydrological Cycle and for Water Management.* 2002 ISBN 1-4020-0444-3
11. N.H. Ravindranath and J.A. Sathaye: *Climatic Change and Developing Countries.* 2002 ISBN 1-4020-0104-5; Pb 1-4020-0771-X
12. E.O. Odada and D.O. Olaga: *The East African Great Lakes: Limnology, Palaeolimnology and Biodiversity.* 2002 ISBN 1-4020-0772-8
13. F.S. Marzano and G. Visconti: *Remote Sensing of Atmosphere and Ocean from Space: Models, Instruments and Techniques.* 2002 ISBN 1-4020-0943-7
14. F.-K. Holtmeier: *Mountain Timberlines. Ecology, Patchiness, and Dynamics.* 2003 ISBN 1-4020-1356-6
15. H.F. Diaz (ed.): *Climate Variability and Change in High Elevation Regions: Past, Present & Future.* 2003 ISBN 1-4020-1386-8
16. H.F. Diaz and B.J. Morehouse (eds.): *Climate and Water: Transboundary Challenges in the Americas.* 2003 ISBN 1-4020-1529-1
17. A.V. Parisi, J. Sabburg and M.G. Kimlin: *Scattered and Filtered Solar UV Measurements.* 2004 ISBN 1-4020-1819-3
18. C. Granier, P. Artaxo and C.E. Reeves (eds.): *Emissions of Atmospheric Trace Compounds.* 2004 ISBN 1-4020-2166-6
19. M. Beniston: *Climatic Change and its Impacts.* An Overview Focusing on Switzerland. 2004 ISBN 1-4020-2345-6
20. J.D. Unruh, M.S. Krol and N. Kliot (eds.): *Environmental Change and its Implications for Population Migration.* 2004 ISBN 1-4020-2868-7
21. H.F. Diaz and R.S. Bradley (eds.): *The Hadley Circulation: Present, Past and Future.* 2004 ISBN 1-4020-2943-8
22. A. Haurie and L. Viguier (eds.): *The Coupling of Climate and Economic Dynamics.* Essays on Integrated Assessment. 2005 ISBN 1-4020-3424-5

Advances in Global Change Research

23. U.M. Huber, H.K.M. Bugmann and M.A. Reasoner (eds.): *Global Change and Mountain Regions.* An Overview of Current Knowledge. 2005
ISBN 1-4020-3506-3; Pb 1-4020-3507-1
24. A.A. Chukhlantsev: *Microwave Radiometry of Vegetation Canopies.* 2006
ISBN 1-4020-4681-2
25. J. McBeath, J. Rosenberg: *Comparative Environmental Politics.* 2006
ISBN 1-4020-4762-2
26. M.E. Ibarrarán and R. Boyd: *Hacia el Futuro.* Energy, Economics, and the Environment in 21st Century Mexico. 2006 ISBN 1-4020-4770-3
27. N.J. Rosenberg: *A Biomass Future for the North American Great Plains: Toward Sustainable Land Use and Mitigation of Greenhouse Warming.* 2007
ISBN 1-4020-5600-1
28. V. Levizzani, P. Bauer, F.J. Turk: *Measuring Precipitation from Space.* EURAIMSAT and the Future. 2007 ISBN 1-4020-5834-9